ELEPHANTS
ARE NOT PICKED
FROM TREES

ELEPHANTS
ARE NOT PICKED
FROM TREES

Animal Biographies
in Gothenburg
Natural History Museum

Liv Emma Thorsen

Aarhus University Press |

Elephants Are Not Picked From Trees
© The author and Aarhus University Press 2014
Layout and cover: Jørgen Sparre
Cover illustrations:
The African elephant and the walrus tusks
in Gothenburg Natural History Museum
Cover photographs: Anders Larsson, GNM
Printed by Narayana Press, Denmark, 2014
ISBN 978 87 7124 212 6

Aarhus University Press
www.unipress.dk

Aarhus
Langelandsgade 177
8200 Aarhus N
Denmark

International distributors:

Gazelle Book Services Ltd.
White Cross Mills
Hightown, Lancaster, LA1 4XS
United Kingdom
www.gazellebookservices.co.uk

ISD
70 Enterprise Drive, Suite 2
Bristol, CT 06010
USA
www.isdistribution.com

Published with the support of
The Research Council of Norway

Content

Aknowledgements

My sincere thanks go first to Göteborgs Naturhistoriska Museum, where I have been received with generosity, goodwill, support and a hearty "Welcome home!" whenever I have returned for another study visit. I would in particular like to thank Birgitta Hansson. Without her thorough and deep knowledge of the museum archive this book could not have been written. Friederike Johansson has helped me to solve the puzzle of the elephant tusks, and she has stimulated my interest in skeleton material. Thanks also to Åsa Holmberg for letting me have unrestricted access to the large museum collection of photos, and for always letting me have excellent office space when working in the museum. All credit for the new photographs in the book goes to Anders Larsson. Christel Johnsson, Monica Silfverstolpe, and Thomas Gütebier have taught me about taxidermy. In the final stages of the book Per Lekholm and Eva Andreasson have been helpful beyond words.

I would also like to extend my thanks to Världskulturmuseet for their assistance, as well as to Sjöfartsmuseet Akvariet, and Trädgårds-föreningen for their help in finding material about apes in Gothenburg.

This book is a result of the project "Animals as Things and Animals as Signs", funded by The Research Council of Norway and hosted by the Department of Culture Studies and Oriental Languages,

University of Oslo, 2008–2012. Thank you to Brita Brenna, Adam Dodd, Guro Flinterud, Henry McGhie, Karen Rader, Brian Ogilvie, and Nigel Rothfels for constructive input and comments. Adam Dodd and John Anthony have kindly helped me with linguistic issues. The road to publication has been long, but never boring!

A Museum and its Animals

This is a book about stuffed animals, not about taxidermy *per se*, but rather about the biographies and collection histories of four mammals on display in the Gothenburg Natural History Museum: a gorilla, a Tonkean macaque, a walrus, and an African elephant.[1] These animals have been selected because they exemplify four different routes animal bodies have followed into the museum's collections. The gorilla was purchased from Rowland Ward in London, a firm that produced and sold mounted animals to museums and private buyers. The Tonkean macaque had been the museum's mascot during the interwar years, and was known by museum visitors as Monjet the monkey. The walrus, which lost its way and ended up in the small islands and skerries off the coast of Gothenburg, shows that animals may end up as natural history specimens, because they have been the victims of accidents. The African elephant was killed to satisfy a taxidermist's dream of once being allowed to stuff and mount the largest land mammal on earth.

The overriding question in this context is to ask which connections are broken and which are established when an animal is included in a natural history collection. Where did the animal come from? How and by whom was it moved from its habitat to the museum? How is an individual animal transformed into a museum artefact and a scientific object? What may the collecting history of an animal reveal about the importance of animals in history and society? Donna Haraway has stated that behind any representation of an animal, whether it is a stuffed animal, a sculpture or a photograph, there is a multiplicity of objects and encounters between humans and animals. These objects and encounters are the sources for writing new biographies, where the individual animal is placed in historical and cultural frameworks (Haraway 1989: 27).[2] A similar understanding of the interpretation potential of material objects can be found in Lorraine Daston's *Things That Talk. Object Lessons from Art and Sciences*, where she claims that certain objects may "helpfully epitomize and concentrate relationships that cohere without being logical in the strict sense" (Daston 2004: 20). The biographies of the four selected animals raise questions as to whose answers are essential to our understanding of the ways in which a location and a building are transformed into a specific "beastly place", into a natural history museum.[3]

Various groups in and outside Sweden have been involved in the establishment of the Gothenburg Natural History Museum's natural history collections: Museum employees, private persons, hunters, explorers, traders, and sailors. Writing the biography of a natural history specimen means that one has to recontextualize the process historically, socially, and culturally, both within the scope of the "afterlives" of the animals in the museum, and within the limitations inherent in the source situation, in order to give the animals a life before death.[4] The biographies are based almost exclusively on sources found in the Gothenburg Natural History Museum: the labels in the exhibition, the collection registers, the museum photo archive, the museum correspondence archive, board minutes, work records, invoices, annual reports, anniversary papers, newspaper articles – and the animal specimen itself. Thus the book also exemplifies how the archives in a natural history museum may contribute to writing cultural history.

Working with one animal has revealed connections to other animals which have hence been drawn into the text. This applies to the baby elephant, and the hide collection which came to light during the analysis of the African elephant, and it applies to the monkeys displayed together with the Tonkean macaque and the gorilla. During the fieldwork in the museum I have regularly put my writing aside and walked over to inspect the display of the animal behind the glass yet again. Alternating between writing and seeing, moving between text and material objects, has led me to new questions that have helped me manoeuvre the animals out of their glass cabinets and the museum building. Methodologically, this may be described as shifting the animals back in time, to other locations in the exhibition rooms, to the taxidermy workshop and the store rooms, and from the red brick building in Slottsskogen back to their places of origin.

A Brief Presentation of Gothenburg Natural History Museum

Gothenburg Natural History Museum is the second largest natural history museum in Sweden, which today houses a collection of around 10 million animals, with 100,000 of these being fragments of vertebrates.

The history of the museum dates back to 31 October, 1833, when "Kung-liga Vetenskaps- och Vitterhetssamhället", the Royal Society of Arts and Sciences in Gothenburg, met and decided to establish a museum to house the natural objects that had been collected in the county of Gothenburg.[5] An underlying wish behind the decision was to systematize and expand the collections (Fåhræus 1983: 15–16). From 1848, the natural history collections were initially displayed systematically in four rooms in the building formerly occupied by the Swedish East India Company (1731–1813). In addition to the zoological specimens, mineralogy and ethnographic objects were included (Hedqvist 2009: 67). Inspired by English museums, the natural history museum was expanded with collections from archaeology, art, and handicrafts, and on 20 December, 1861, Göteborgs Museum, the Gothenburg Museum, opened (Fåhræus 1983: 20–21).

The new museum used all of "Ostindiska Huset", the East India House, erected in the city centre in 1762. The natural history museum now became part of the Gothenburg Museum, and the exhibitions were placed on the upper floor of the west wing of the building (Hedqvist 2009: 68). The zoological specimens were on display in a room with a gallery and skylights (Hedqvist 2009: 69). The specimens were displayed in cabinets attached to the floor, ceiling, and walls, "which has sealed them, ensuring that in this museum one will not encounter the unpleasant odours otherwise endemic to and developed by such collections" (Carlén 1869: 78).[6] On the inside the cabinets were painted a light blue-green, a colour rarely found in nature, which would thus highlight the animals on display. The exterior paint imitated oak. The furnishings were generally simple in order to give prominence to the objects displayed. In the middle of the floor space was a high octagonal glass case, which contained "a small world unto itself";[7] representatives of animal species which were considered to be the most useful or most noxious in Sweden's agriculture, forestry or gardening (Carlén 1869: 78, 79). Based on what may be dimly seen in a lithograph showing the Museum of Natural History in 1865, the case contained gallinaceous birds, ducks, and geese. The smaller birds on the upper shelf may have been a peregrine falcon, northern goshawk, and common raven, which the then curator of the natural history collections, A.W. Malm, considered to be harmful species of birds, and which he accordingly believed should be exterminated (Mathiasson 1983: 26). The attitude of Malm to nature, as a zoologist, was dominated by contemporary thinking that natural resources should be exploited to the full to promote the primary industries. On the one hand, he founded Sällskapet Småfoglarnas Vänner, the Association of Friends of Small Birds, on the other hand he encouraged the killing of birds of prey and the draining of wetlands with rich populations of birds (Mathiasson 1983: 31).

The gorilla from Rowland Ward was purchased in 1906, and the mounting of the African elephant was completed in 1952. During this period of time the museum collection of foreign mammals was doubled, from 639 in 1910 to 1183 in 1958 (Mathiasson 1983: 49). The most important event in the history of the Gothenburg Natural History Museum between these two years is the construction of a new building in Slottsskogen. On 8 July, 1923, the year Gothenburg celebrated its 300 year anniversary, the new natural history museum opened its doors

Fig. 1. The skylight room with the natural history exhibitions in the
East India House in 1865. The octagonal case with useful
and noxious birds is at the back of the room.

to the public.[8] The guest list counted more than 300 individuals and its configuration reflects the fact that the natural history collections in Gothenburg were considered to be significant and of scientific interest: The head of the Academy of Science was one of the speakers, and the guests represented Gothenburg's city administration, and the Nordic universities and natural history museums (Hedqvist 2009: 120).

The rich collection of clippings found in the museum's archives from this period shows that there was a keen interest in the museum; new animals in the display collection were especially popular. The many Gothenburg newspapers followed closely the work carried out by the museum, not least the taxidermists' endeavours, and reported on Monjet the monkey and the walrus. The acquisition and the mounting of the African elephant attracted even more attention. The museum issued printed invitations to prominent citizens, when the gorilla and the walrus were installed. Today children and pupils constitute the majority of the visitors. During the period in question, exotic and preferably large stuffed mammals attracted an adult, well-educated, and affluent public to the museum. Rare animals aroused curiosity even as "still life".

Gothenburg is a relatively young city in a Swedish context. It has, however, been the leading trade and maritime city in modern Sweden. A travel guide from 1869 says:

Admittedly, Gothenburg lacks the glamour of historical monuments; no Sture fought there for the independence of the mother country; no Christina Gyllenstjerna defended its walls, and no Gustaf Wasa marched victoriously into the city. But Gothenburg shines with a different light, one that produces shields and weapons, which thwart or defeat the most dangerous enemy of our time, pauperism. This light is the mercantile and industrial superiority Gothenburg has conquered. This initiative has contributed to many vital decisions and actions that were important for the city and even the entire country. (Carlén 1869: 3–4)[9]

In the years around 1900 a number of cultural institutions were established in Gothenburg based on donations from affluent citizens (Hedqvist 2009: 92). Similarly, representatives of the city's trade and industry patricians were on the board of the Gothenburg Museum (Hedqvist 2009: 108), and they donated the funds for special purchases when the annual budget for the Zoological Department was insufficient. These men were also "skilled amateur (naturalists) who constituted a circle associated with the museum" (Hedquist 2007: 185).[10] The gorilla and the elephant were both added to the collections thanks to donations from private benefactors.

Over the last few decades several well-known natural history museums have restructured their exhibitions. Some examples include Musée national d'histoire naturelle in Paris, Naturhistoriska Centralmuseet (The Finnish Museum of Natural History), The Manchester Museum and Naturhistoriska Riksmuseet (The Swedish Museum of Natural History). Old exhibitions have been reorganized or dismantled to provide room for updated presentations of nature. The systematic displays of natural history are being replaced by presentations of ecological relationships and biological research. Fortunately for this study, very few changes have been made to Valsalen, the Whale Hall, and

Däggdjurssalen, the Mammal Hall, after the animals discussed here were put on display. The display cases have become fuller, as new animals arrived, but once the animals were ensconced behind glass, they were frozen in time and space. The exception is the walrus, which has been moved several times inside the museum.

The museologist Eric Hedqvist has pointed out that the exhibitions offered to the public in the new museum were outdated: "The exhibitions were in general based on a descriptive tradition and did not visualize the prerogative of modern science, the distinctive causal connections". (Hedqvist 2007: 188, 189).[11] The first known habitat dioramas were mounted in Sweden by zoologist, taxidermist, and hunter Gustaf Kolthoff's (1845–1913) biological museums (Wonders 1993: 46–71). The technique was developed by Olof Gylling (1870–1929), taxidermist at Malmö Museum. Gylling prepared five exquisite dioramas in the new natural history museum in Gothenburg (Wonders 1993: 76–83).[12] Apart from these and the biological bird groups, mammals were displayed systematically, emphasizing their morphology and each animal had its own text on a label. This indicates that many of the stuffed mammals are old. A number of them were collected around 1900 before the major move from the East India House to Slottsskogen.

Moving the natural history collections to the new building in Slottsskogen separated zoology and geology from the cultural history collections in the Gothenburg Museum, as they were placed in a building especially designed for the collecting, researching, and presenting nature. This was not unique to Gothenburg, but a process that can be traced back to the end of the 1800s and the first decades of the twentieth

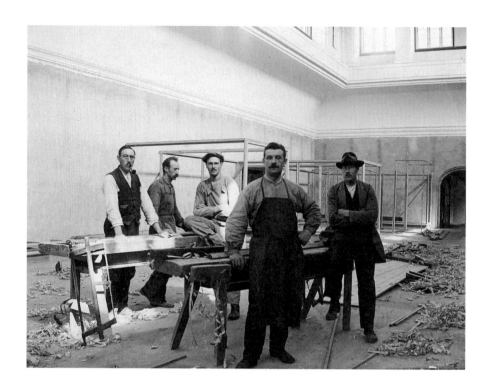

Fig. 2. Making space for animals.
The empty Mammal Hall in the new museum in Slottsskogen.
Carpenters constructing glass cabinets, 1920.

GNM_5662_2.

century, when specially designed buildings were erected in Europe and the USA to house natural history collections and research activities connected to the collections. In the words of architecture historian Carla Yanni, before this time "the pursuit of natural knowledge ... [took place] in gentlemen's houses" (Yanni 2005: 1). With its new building in Slottsskogen, Gothenburg joined the ranks of cities with modern natural history museums, whereof one of the first, largest and most famous

Fig. 3. A bongo being moved to the new museum building,
probably in the autumn of 1921. Two hides of bongo antelopes and an okapi
hide were bought by the museum in 1912 from a Swedish ship's captain.
The hides were presumably smuggled and exported illegally,
but since the animals were already dead, there was no harm in buying them
(Jägerskiöld 1943: 381).

GNM 5345_29.

was the Natural History Museum in London, designed by the architect
Alfred Waterhouse and completed in 1880 (Stearn [1981] 2001; Yanni
2005: 111–147). On the one hand, the knowledge about nature was materialized in the form of buildings and objects, which like today, drew

Fig. 4. Paul Henrici in the Mammal Hall photographed during the move into the new museum building in the autumn of 1922. Jägerskiöld writes about Henrici: "He has a fine, almost fastidious taste when it comes to museum affairs. It was quite appropriate at our installation in 1923"[3] (Jägerskiöld 1943: 400).

a large public to the natural history museums. On the other hand, this led to a clear division between museums of cultural history and natural history, between culture and nature.

The process of moving the collections from the East India House to the new building in Slottsskogen took several years. The exhibition rooms were also to be made ready for the public, which is to say that they were to be filled with specimens and mounted animals. The museum database of photographs has pictures showing carpenters in an empty Mammal Hall. Another picture shows a row of stuffed animals "waiting" to be inserted in display cabinets. While the moving process went on, Gothenburg residents were able to witness domestic and exotic mammals being carted and driven from downtown up to Slottsskogen, which must have been both an amazing and marvellous sight. The animals were moved from one building to another, but which journeys had they undertaken before being added to the collection, catalogued, stored, and mounted?

A Culture of Hunters

"Hunting and shooting have helped in my work as a zoologist and museum man. It has also given me spiritual calm and bodily refreshment to the highest extent",[14] Leonard Jägerskiöld writes in his autobiography *Upplevt och uppnått* (Jägerskiöld 1943: 455). His full name was Axel Krister Edvard Leonard Jägerskiöld (1867–1945). Jägerskiöld was hired as "intendant för de naturhistoriske samlingarna", Head of the museum, on 1 June, 1904, and was in charge of the Gothenburg Natural History Museum until 1 May, 1937. Thus Jägerskiöld is mentioned in the biographies of all the four animals examined here, even if the elephant was collected for the museum under his successor Orvar Nybelin (1892–1982).

It was not least Jägerskiöld's energetic efforts that resulted in a new museum building. Jägerskiöld wanted to "move nature into the museum and to give an idea about how the animals lived, how they spread and their characteristics" (Jägerskiöld quoted in Orrhage 1983: 35).[15] One means of moving nature was hunting. An extensive chapter has been devoted to hunting in Jägerskiöld's memoirs, and they reveal a close connection between being a professional in the museum world, whether as a zoologist or taxidermist, and being a hunter. Jägerskiöld himself and his employees contributed actively to moving nature into the museum. As Jägerskiöld puts it: "One might almost say that the entire staff of the Natural History Museum joined in the hunt".[16] He liked to hunt seal in the skerries in Bohuslen county with Stin-Olle and his son Utter-Anders who lived on the island of Rörö. Olle got the blubber, Jägerskiöld bought the skin, the museum received intestinal parasites and other vermin, and the meat was thrown into the sea (Jägerskiöld 1943: 467). Each summer the museum staff would be collecting birds, eggs, and nests, and several were prepared and displayed to the public as "biological bird groups".[17] Jägerskiöld's preferred hunting companion among the museum staff was Associate Professor Paul Henrici (1919–1948), but he also hunted with the museum taxidermist Hilmer Skoog (1870–1927), who was self-taught, and Skoog's successor David Sjölander (1886–1954).[18] As Sweden was a country with strict class divisions, these hunting trips must have contributed to evening out the social differences within the staff. As Jägerskiöld puts it in his memoirs, the museum staff was a closed circle, "one might almost say a family" (Jägerskiöld 1943: 402).[19]

Knowledge about nature and animals was anchored in the interest and pleasure of being outdoors with a gun. Hunting could be combined with collecting and scientific observations, and the museum zoologists and taxidermists participated in this "hunting-zoology". Jägerskiöld took it for granted that a taxidermist would also be a hunter, not only in order to collect animals for the exhibitions, but also to gain knowledge about animal ethology, thus becoming more professional at work. Jägerskiöld, for example, encouraged Skoog to hunt, so that he would see how animals moved in nature (Jägerskiöld 1943: 393). The dead animal was a requirement for the museum work, and knowledge about nature, accumulated through numerous hunting trips, was invested in the collections and in the displays. Practice from the hunting fields was converted into the systematic displays in the cabinets.

In her study of the history of the diorama in Sweden, and as one of the explanations as to why the Swedes began making habitat dioramas so early, the art historian Karen Wonders points to the tradition in Sweden of hunter-naturalists (Wonders 1993: 89).[20] She traces the importance of hunting for the collecting of specimens, and conversely, the importance of the demand for specimens for hunting, back to Sweden's first Professor of Zoology, Sven Nilsson (1787–1883). Nilsson believed that "zoological research should be conducted in *living nature*, rather than in natural history collections of dead objects" (Wonders 1993: 90). Like the hunter, the zoologist should be familiar with the animals in nature. At a later stage Nilsson developed an interest in methods to preserve zoological specimens, and in 1828 he became the curator of the natural history collections of the Royal Swedish Academy of Science and the newly founded Swedish Museum of Natural History in Stockholm. In 1832 Nilsson was appointed Professor of Zoology in

Lund, where he also renewed the university's zoological collections. The links between zoology and hunting are also demonstrated by Nilsson's active participation in the founding of the Swedish Association of Hunters,[21] and the hunters were encouraged to support zoological research and donate game to museum collections (Wonders 1983: 91).

In Sweden, as mentioned above, the natural history dioramas were first introduced in Gustaf Koltoff's biological museum, and Kolthoff continued to unite hunting and zoology, a practice that Nilsson had introduced (Wonders 1983: 91). He took the initiative to build three biological museums of which the best known is the Biological Museum at Djurgården in Stockholm, which opened in 1893. Kolthoff was a taxidermy student at the National Museum of Natural History (Svenska Riksmuseet) in Stockholm in the 1860s, and it has been claimed that he developed the idea for the dioramas as a way of counter-balancing the assemblage of stuffed animals in the semi-dark rooms as seen in the exhibitions of the National Museum of Natural History. Gunnar Brusewitz writes: "The zoological museum had been transformed into a kind of scientific mausoleum, which must have appeared as a mockery of nature for a sensitive aesthete such as Gustaf Kolthoff" (Brusewitz 1993: 11).[22]

Kolthoff's biological museum could be called an "ecological museum". The animals must be displayed in their correct environment, in their habitats, where everything must be reconstructed so that the display would stimulate the imagination, and also be beautiful and instructive. What is unique about the biological museum's exhibitions of stuffed animals, and what distinguishes it from many other natural

Fig. 5. The culture of hunter-zoologists on display in Ostindiska Huset.

PHOTO: AXEL LINDAHL. GNM_5599_8.

history museums, is that it activates the visitor in a different manner. Most exhibitions tell the visitor in detail what can be seen behind the glass. Kolthoff's exhibitions offer information for each diorama with respect to what it is possible to see behind the glass, and then it is up to the visitor to find the animal or the bird. Thus visitors were taught how to discover and see animal life in nature, the woodpecker's hole, the willow warbler's nest – knowledge that could then be brought into nature (Brusewitz 1993: 19). Kolthoff and Jägerskiöld were close friends who shared an enthusiasm for hunting and an interest in presenting zoology to a broad public via museum exhibitions. In spite of this, the museum exhibitions they made were very different, Kolthoff introduced the habitat diorama, while Jägerskiöld stuck to the well-known systematic presentation of the species, void of pointers to the animals' habitats.

Transformations

The biographies of the four animals show that the live animal is relocated from its habitat to the museum through stages and complex transformations. They confirm the historian Sam Alberti's statement that the biographies of zoological specimens are *"material knowledge in transit, bringing experiences of nature with them to different sites and audiences"* (Alberti 2011: 4). The historian Reviel Netz maintains that the core of history concerns individuals of flesh and blood interacting in a spatial materiality: "History is not at all abstract: it is a matter of flesh-and-blood individuals interacting in material space." He continues: "History takes place when flesh moves inside space; it is thus, among other things, about the biology of flesh – as well as about the topology of space." And individuals encountering and interacting are humans as well as animals: "Thus history is embodied – and not only inside human bodies but in the bodies of all species" (Netz 2004: 229). The transformations of the animals discussed in this book were effected through a number of moves and border crossings, outside as well as inside the museum, where the major transformation of the animal body is from life to death, followed by the modelling of a new body, adapted to the animal's hide. Scars, bones fused after fractures, bullet injuries and bullet holes in hides and skeletons show that the history is physically etched into the animal bodies.

The transformation of an animal body into a specimen means preserving it for the future, placing the conserved object in a room that can optimally slow down entropy, and give it new meaning by entering it in the museum registers. The remains of what once was a walrus, a Tonkean macaque, and an African elephant can now be found in several locations in the museum: The mounted skins in the Mammal Hall, the skeletons in the Bone Cellar, and the intestines in the wet store room. Alberti maintains that objects in the natural history museum must be "impartial", that is to say they cannot be value-charged if they are to function as data (Alberti 2008: 81). The fact that mounted animals are handcrafted products, must be hidden so that the museum is not undermined as an institution that explores nature, Alberti asserts. The core of the process is "the very act of removal" (Alberti 2008: 82). An animal on display in a natural history exhibition has been "threshed" clean of meanings that place the animal in culture and society.[23] Threshing grain means to remove the husks that cover the seed to separate the nourishing core. Applying this as a metaphor of how animal bodies have been treated in the natural history museum, implies that in a museum context there must be a genuine animal, a being which is pure nature without any cultural connotations.

The taxidermists Hilmer Skoog and David Sjölander transformed the walrus, the Tonkean macaque, and the elephant into excellent handcrafted products. Skoog shaped the walrus and Sjölander was the master who did the monkey and the elephant. They gave the animals new bodies, after skinning and skeletonizing them and placing the intestines in alcohol. A taxidermist makes what we in vernacular speech call stuffed animals. Under the skilled hands of a taxidermist a flat hide becomes a three-dimensional animal. The term taxidermy comes from the Greek, being composed of 'taxis' which means movement, and 'derma' which means skin, so the combination of the two creates the meaning of moving or arranging the skin or the hide in the sense of manipulating it so that it will have the same shape the animal had, when it was alive (Morris 2010: 8). The taxidermist mounts the hide on a body designed according to measurements and photos of the animal, as opposed to the old way where the hide was stuffed or upholstered with straw, paper or another material, which resulted in lumpy and lifeless forms. "To naturalize" is another verb used about the mounting of animals, and it expresses the purpose of the taxidermist's work: To achieve a result which shows the animal in its natural state. In the words of Alberti: "Mounts are intended to be 'resurrections', as close to life as possible" (Alberti 2008: 81).

Taxidermy was developed into a refined craft in the course of the second half of the 1800s and first half of the 1900s. Knowledge of the ways in which feather and skin could best be prepared and preserved was one of the requirements for successful presentation and research, and taxidermy became an important aid in the building of zoological collections in the many new natural history museums established in the West during this time (Star 1992). White Europeans and Americans went on hunting expeditions to Africa and Asia, returning with trophies. Moreover, they had hides and horns to sell to such buyers as the natural history museums. This created a market for the taxidermists' products. The most talented taxidermists knew how to exploit this segment by displaying the specimens at world exhibitions and other international exhibitions. The Great Exhibition in London 1851 displayed trophies of animals from Great Britain and from the British colonies (Ritvo 1987: 249). The English taxidermist Rowland Ward (1848–1912) gained international recognition by participating in a number of international exhibitions (Morris 2003: 33). The breakthrough for famous American taxidermist Carl Akeley (1864–1926) came at the World's Colombian Exhibition in Chicago in 1891, where he showed three mounted broncos (Bodry-Sanders 1998: 38).

Akeley first modelled an anatomically correct body of clay based on exact measurements, plaster casts of the head and other parts of the animal, as well as a small plaster model showing what the finished product should look like. Thereafter the clay body was covered with plaster. After the plaster had hardened, it was split and then put together again as a negative form. Using the negative form, Akeley cast a papier-mâché manikin onto which he mounted the hide. This method, where the hide is mounted over and attached to a modelled body, is called the dermoplastic technique and was developed by the German Philipp Leopold Martin (1815–1885). The disadvantage of Martin's method, where the hide was attached to a modelled body covered by a layer of moist clay, was that the specimen became heavy, and that the skin would shrink and craze as the clay dried out (Wonders 1993: 40– 41).

Akeley made two great contributions to the dermoplastic technique. The first was the making of an anatomical model of clay, which made adjustments easy and allowed for the reuse of the modelling clay and racks used in the construction. Secondly and most important innovation was the introduction of the technique of using negative plaster forms in several sections, which made it easy to cast the manikin in a light material. Today the manikin is cast using polyurethane foam.[24] Skoog and Sjölander both used the dermoplastic method. Skoog worked according to Martin's method by making a model to which he nailed and glued the hide. Sjölander, on the other hand, used Akeley's method, as did his successor Björn Wennerberg (1910–1978) (Setterberg 1989: 24–25).

In her article "The Matter and Meaning of Museum Taxidermy", the art historian Rachel Poliquin maintains that stuffed animals are opaque and problematic to grasp: "If taxidermied animals were easy to read, the process of looking at taxidermy would hardly be worth the effort" (Poliquin 2008: 133). What distinguishes stuffed animals from models made from glass and wax is that while the latter are similes, i.e. they are imitations of originals, the stuffed animals have skin and fur, which once belonged to a live animal, pretend to be the real thing: "This uncanny animal-thingness has the power to provoke, to edify, and even to undermine the validity of its own existence" (Poliquin 2008: 127). I share Poliquin's view that it is the disquiet, and I would add melancholy feelings these "animal-things" evoke, as well as the fact

Fig. 6. From the mounting of the African elephant, which was mounted according to Akeley's method. The clay body is covered by a layer of plaster sectioned into pieces. These pieces were used for casting the pieces of the manikin in plaster. The manikin plaster pieces were joined together with plaster and glue, and reinforced with jute weave. The elephant's leg was strengthened with steel poles.

PHOTO: DAVID SJÖLANDER. GNM4576_10.

that they cannot be placed in well-defined categories, that pique one's curiosity.

In the essay "I tingenes mellomrike" (In the In-between Realm of Things) the philosopher Dag T. Andersson shares similar thoughts, writing that it "may prove to be productive to apply the tension embedded in a common concept of things, where the thing is thought of as an entity located between nature and culture, and where it is not always completely clear whether the emphasis should fall on the side of nature or of culture" (Andersson 2001: 132).[25] Referring to the author Geoffrey Grigson, Andersson suggests that dolls are the most secretive objects "in the boundary between a human being and objects" (Andersson 2001: 134).[26] Andersson points out that all things ascribed to civilization and culture contain an element of nature (Andersson 2001: 133). Mounted animals remind us of dolls. The fur is mounted on a modelled body or manikin and the eyes are made of glass. And as with dolls, mounted animals have a status that cannot be fully comprehended. In the Gothenburg Natural History Museum, the mounted species exemplars are arranged taxonomically. They have a didactic purpose by acting as three-dimensional natural history illustrations. Each animal is furnished with a label. As I intend to show, the text of the label moves the animal out of its category, defined by natural science, and into an "in-between realm of things".

In recent years, researchers, particularly in the humanities, have begun to study taxidermy and particularly taxidermist products. This is due to the fact that the humanities have found renewed interest in the materiality and materializations of culture, in museology, and in the particular materiality found in the collections and exhibitions of the natural history museums (Alberti 2005 and 2008, Damsholt, Simensen, and Mordhorst 2009).[27] It is therefore not correct when Rowland Ward's biographer, P.A. Morris, claims that taxidermy "is an aspect of social and natural history that has been largely forgotten, in a form of collective amnesia, or swept under the carpet like some kind of national disgrace" (Morris 2010: 5). In fact the opposite is true. Mounted animals are life without breath, and it is this property that has made artists, researchers, and journalists speak about them, display them and write about them (Alberti 2011, Eastoe 2012, Madden 2011, Milgrom 2010, Poliquin 2012, Rothfels 2013, Snæbjörnsdóttir and Wilson 2006, Turner 2013).[28]

The mounted animals on display in cabinets and dioramas are standardized objects made to illustrate and represent nature. However, if objects could talk, these objects speak to us in low voices about the historical individual, about the species in science, and about the animal's representation in cultural history; they speak about both the animals' natural history and their unnatural history.[29]

This ambiguity has inspired researchers to write biographies based on the remains of dead animals. I would like to highlight Christopher Cokinos' deeply engaging study *Hope is the Thing with Feathers*, about the last live individuals of six now extinct North American bird species, the Carolina parakeet, the ivory-billed woodpecker, the heath hen, the passenger pigeon, the Labrador duck, and the great auk (Cokinos 2009 [2000]). Cultural history monographs have been written about journeys famous exotic and historical animals were forced to make to and in Europe (Allin 1998, Bedini 1997, Ridley 2005

[2004]). In the history of zoos, certain animals – usually large mammals – have become famous. A late modern example is the polar bear called Knut in the Berlin Zoo who died in 2011 only five years old. In *A Polyphonic Polar Bear* Guro Flinterud highlights the ambiguous meanings that various groups linked to this animal celebrity (Flinterud 2013). As Alberti points out, historians have primarily been interested in conspicous and iconic mammals (Alberti 2011: 10). *The Afterlives of Animals* (Alberti 2011) is a collection of animal biographies based on remains in natural history museum collections. The anthology includes two articles, where the authors take their material point of departure in a flat hide, exemplifying the thousands and thousands of anonymous hides conserved, catalogued, and hidden in museum store rooms all over the world (Everest 2011, Patchett, Foster, and Lorimer 2011).

Four Animal Bodies

A lowland gorilla, a Tonkean macaque, a walrus, and an African elephant have given bodies to the four main chapters of the book. Animals have always functioned as useful containers for human ideas. Museums have had to take into consideration what their visitors have been willing to accept seeing in the displays. The characteristics a certain species of animal has been assigned have changed over time, a fact emphasized by the collection histories of the gorilla and the walrus. The exhibitions of the natural history museums are based on the general acceptance of the fact that animal bodies can be stuffed and exhibited. There were no protests against the preparation of these animals, actually the process was followed with keen interest by the media. However, in the twenty-first century, we have other ideas and opinions about animals than what we had a century ago, when Leonard Jägerskiöld proudly displayed a stuffed gorilla, an orangutan, and a chimpanzee to the public. A radical example of this change in attitudes is the initiative by American animal rights activists to change the legal status of chimpanzees into legal "persons". What will the public reaction be if the chimpanzee's legal status is changed from being a thing into a "person"? Would taxidermied bodies of chimpanzees and the other great apes then continue to be exhibited in museum cases?

The natural history museum is filled with animal remains that were once parts of a living organism. A general problem of the museums is constituted by the broken relations between the objects in the collections and the contexts they were once part of, and in this particular case the lost connections between the once living animal and the specimen. The truth about taxidermy, Poliquin says, is that "it is simultaneously a representation and a presentation of animal form" (Poliquin 2011: 107). The intention of the museum is to display specimens of species where, reasonably enough, specimen-specific characteristics are highlighted instead of individual features. The gorilla, the macaque, the walrus, and the elephant are didactic objects, and good craftsmanship has also turned some of them into aesthetic objects. By reconnecting the broken ties between the concrete animal form and the living animal it once was, the stuffed animal is sent into the "in-between realm of things" in a room where the boundaries drawn by the natural history museum between nature and culture, and between specimen and individual, are destabilized. This opens a different and demanding narrative about the bodies in the glass cases. When the animals become cultural history actors, they also become provocateurs.

In the four main chapters the mounted animal hides are linked with networks that were active in the collection of natural specimens undertaken by the museums. In all the chapters I have endeavoured to highlight the individual animal that "donated" its skin to the museum, the natural properties of the species, as well as the attending cultural issues.

Chapter 1 – *Commercialized: The Gorilla from Rowland Ward* – places the gorilla in a network of hunters, taxidermists, and trade around 1900, which aimed to provide the many new natural history museums with specimens. In these transactions the individual animal disappeared. The text focuses on the taxidermists' as well as the current interpretations of the gorilla in that day and age. Chapter 2 – *Captured: Monjet the Monkey* – is the story of a small Tonkean macaque which left many traces while still alive, but which lost her individuality, when she was naturalized and placed in a cabinet with other monkeys. Monjet's biography is also a contribution to the cultural history of the monkey in the West. Chapter 3 – *Stranded: The Walrus from Rörön* – follows the long swim of a male walrus from Eastern Greenland to the Swedish island of Rörön, and discusses the transformation from individual to specimen, emphasizing what happens when people encounter unknown, large, and ugly animals. Chapter 4 – *Collected: The African Elephant* – highlights the work inherent in moving an elephant's body from Portuguese West Africa to the Mammal Hall to recreate it in the

Gothenburg Natural History Museum. Moreover, the chapter looks at the narratives lost in the transformation. The animal biographies are ordered chronologically.

The individual collection histories of the animals link them to the intense hunt in the early 1900s for exotic animals for the zoos and natural history museums, as well as to the demand for the good specimens. Gothenburg Natural History Museum was, on the one hand, a museum on the outskirts of Europe compared to the prestigious natural history museums further south and in the major American cities. On the other hand, the museum management traded within an international network and acquired the material for the collections through the same channels as the large European natural history museums.

The biographies also show that several of those who provided and preserved zoological material for the museum collections saw no contradiction between showing respect for live animals, killing animals, and preserving mounted animals. The museum exhibitions also had to be beautiful, and by displaying good specimens in well-arranged cases, the visit to the museum would become an aesthetic experience, as well as an instructive one. The gorilla, the walrus, and the African elephant all in turn became public attractions. The Mammal Hall in Gothenburg Natural History Museum still displays the gorilla, Monjet, and the elephant, and these days the exhibition hall is as an example of how a beautiful exhibition of mammals from local and far corners of the world was presented in the early 20th century.

Notes

1   The Swedish name of the museum is Göteborg Naturhistoriska Museum.

2   "Behind every mounted animal, bronze sculpture, or photograph lies a profusion of objects and social interactions among people and other animals, which can be recomposed to tell a biography embracing major themes for twentieth-century United States" (Haraway 1989: 27).

3   The expression "a beastly place" refers to the title of C. Philo's & C. Wilbert's collection *Animal Spaces, Beastly Places* (2000).

4   See the collection Alberti (2011), *The Afterlives of Animals*, about natural history specimens in British museums.

5   A.W. Malm, curator for the natural history collections 1879–1882, assumed that the oldest specimens were added after 1721, and would have come from the old high school collection (Carlén 1869: 73). The description of the Natural History Museum in Carlén's guide *Göteborg. Beskrifning öfver staden och dess närmaste omgifningar. Ny handbok för resande* is written by Malm.

6   "… hvarigenom en täthet vunnits, som gör, att man i detta museum ej mötes af den skarpa lukt, som annars tillhör och utvecklas af sådana samlingar" (Carlén 1869: 78).

7   "… en liten verld (sic) för sig" (Carlén 1869: 79).

8   Botaniska Trädgården [The Botanical Gardens] also opened in 1923. From 1916 to 1933 an impressive series of new museum buildings was constructed in Gothenburg, several with collections that used to be part of the Gothenburg Museum, such as Rösska Museet [The Rösska Museum] (1916), Göteborgs Konstmuseum [The Gothenburg Museum of Art] (1920) and Göteborgs Sjöfartsmuseum [The Gothenburg Maritime Museum] (1933).

9   "Visserligen saknar Göteborg glansen af historiska minnen; ingen Sture har der kämpat för fosterlandets sjelfständighet; ingen Christina Gyllenstjerna har försvarat dess murar, och ingen Gustaf Wasa har der hållit sitt segerinntåg. Men Göteborg äger en annan glans, med hvilken det smider sköldar och vapen, som afvända eller besegra vår tids farligaste fiende, pauperismen. Denna glans är den merkantila och industriella öfverlägsenhet, som Göteborg tillkämpat sig, och det initiativ det gifvit till flere för staden och hela landet vigtiga beslut och åtgöranden" (Carlén 1869: 3).

10 "… kunniga amatörer som bildade en krets kring museet" (Hedquist 2007: 185).

11 "Utställningen var i huvudsak baserad på naturhistorisk beskrivande tradition och visade inte de för modern naturvetenskap utmärkande orsakssambanden …" (Hedqvist 2007: 188, 189).

12 The dioramas were funded by a donation of 50,000 Swedish crowns from businessman and Gothenburg resident Gustaf Werner. Wonders claims that without this donation the museum would not have had the funds to make the diorama (Wonders 1993: 77–78).

13 "Han har en fin, nästen grätten smak i museala ting. Den kom väl till pass vid vår installation 1923" (Jägerskiöld 1943: 400).

14 "Jakten och skyttet ha hjälpt mig i mitt yrke som zoolog och museimann. De ha dessutom skänkt andlig vila och kroppslig vederkvickelse i rikaste mått" (Jägerskiöld 1943: 455).

15 "… flytta in naturen i museet og ge en föreställning om djurens levnadssätt, utbredning och byggnad" (Orrhage 1983: 35).

16 "Man kan nästan säga att hela Naturhistoriska Museets stab var med och jagade" (Jägerskiöld 1943: 486).

17 "biologiska fågelgrupper".

18 David Sjölanders professional competence is discussed in Chapter 4.

19 "… jag hade när sagt en familj" (Jägerskiöld 1943: 402).

20 "jaktzoologer".

21 Svenska Jägareförbundet.

22 "Det zoologiska museet var förvandlat till ett slags vetenskapeligt mausoleum, som måste ha framstått som en skymf mot naturen för en kännslig estet som Gustaf Kolthoff" (Brusewitz 1993: 11).

23 S. Alberti used this metaphor in a talk given in Oslo, October 2009.

24 The most important methods used today are the Akeley method, the ter Meer method, the Kerz method and the Küsthardt method. E-mail from taxidermist Thomas Gütebier 20.02 2011. On Akeley's method see Gütebier 2011, on dermoplasticity and the German tradition see Gütebier 1995.

25 "Det vil kunne vise seg fruktbart å utnytte den spenning som ligger i et gjengs begrep om ting, der tingen tenkes som en størrelse som har sitt sted mellom natur og kultur og der det ikke alltid er helt klart om det er på natur- eller kultursiden tyngdepunktet skal legges" (Andersson 2001: 132).

26 G. Grigson: *Things*. New York 1978. "… i et grenseland mellom menneske og ting" (Andersson 2001: 134).

27 Alberti 2008 gives an excellent overview of research literature in English which analyses natural history objects and collections.

28 Bryndís Snæbjörnsdóttir and Mark Wilson in cooperation with the Research Council of Norway project "Dyr som ting og dyr som tegn" [Animals as objects and animals as signs] made the art installation "Animal Matters", which was shown in Galleri Sverdrup, the University of Oslo, in the summer of 2012.

29 The expression "animals' unnatural history" was introduced by N. Rothfels (2002).

Gorilla gorilla (Wyman), male

Provenance: Gabon, West Africa

Date of death: Unknown

Death: Unknown

Collected by: Unknown

Mounted by: Rowland Ward Ltd.

Owner: Gothenburg Natural History Museum

Condition: Good

Entry in General Register: 1906-1116

Entry in Register of Foreign Mammals: Ma.ex. 610

Entry in Collectio Anatomica: 2978

Place in the museum: Body displayed in the Mammal Hall

Skull in the Bone Cellar

Location in the museum: Mounted hide in display in the Mammal

Hall; skull in the Bone Cellar

Red-listed: Critically endangered

Commercialized : The Gorilla from Rowland Ward   1

On 3 October, 1906, a telegram was sent from Gothenburg to London:

> Jungle London
> Please send stuffed Gorilla skull and skeleton. Money procured. Letter follows.
>   Jägerskiöld

The order settled a bargain between the Head of the Zoological Collections at Gothenburg Museum, Leonard Jägerskiöld, and the internationally renowned taxidermy firm, Rowland Ward Ltd. The mounted body of the Ward gorilla will in this chapter be examined as both specimen and commodity – one among the thousands of beastly bodies entangled in the animal business around 1900. The gorilla serves as a general example of how museums obtained large, exotic mammals in the early twentieth century, and demonstrates how both the Gothenburg Museum and the gorilla were situated within a network of animals, hunters, taxidermists, businessmen, and museum curators.

It has not been possible to thoroughly trace the gorilla's international journey – from Gabon to Gothenburg, via London. However, the archived correspondence between Leonard Jägerskiöld and Rowland Ward Ltd.,[1] in addition to Jägerskiöld's correspondence with the

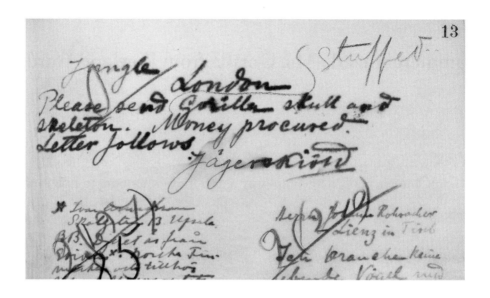

Fig. 7. Jägerskiöld's original text for his telegram to Rowland Ward,
telling that the money for the gorilla had been procured.

PHOTO: ANDERS LARSSON, GNM.

Swiss Georg Albert Girtanner, as well as the photographs of dead and mounted animals sent from Ward's and the German firm J.F.G. Umlauff to the museum, shed some light on how the great apes became objectified and commercialized at the time. The chapter deals with three questions: How did museums acquire and accommodate large simian specimens around 1900? How were the largest ape specimens represented in taxidermy between 1900 and 1927? And finally, what do the poses in which mounted gorillas were positioned tell us about conceptions of the gorilla itself?

London, Hamburg, and Gorillas

Around 1900 London and Hamburg were among the most important mercantile ports in Europe. Living and dead exotic animals arrived at these hubs from the colonies: the living ones continued to zoos, while the hides and skeletons of the dead were transformed and traded by taxidermy companies. Two of the most renowned European taxidermy firms, Rowland Ward Ltd. (1872–1983) and J.F.G. Umlauff (1868–1974), were located in London and in Hamburg, respectively. In 1906 the Gothenburg Museum's Zoological Department patronised both of them. Ward's was strictly a taxidermy firm, while Umlauff procured ethnographic material, biological specimens of humans and animals, and models of 'Naturvölkern' (native peoples), taxidermy accounting for only a portion of their business. The firms' reputations differed, too. Ward's was popular with taxidermy connoisseurs appreciating the company's quality products supplied throughout its hundred years of service. Their customers were primarily big game hunters, private collectors, and museums. Among their most prominent patrons were Prince Philippe, the Duke of Orléans, the English banker and zoologist Lord Walter Rothschild, and the British Museum in London. While the French duke was an avid big game hunter and trophy collector, Lord Rothschild collected specimens for his private natural history museum in Tring. After fifty years of collecting, Lord Rothschild had amassed "the greatest collection of animals ever assembled by one man" (Rothschild 1983: 2). He was also a great benefactor, donating many specimens to the British Museum during his life, and bequeathing his entire museum to the nation upon his death in 1937.[2] Ward's also bought skins to be set up for sale "especially acting as an agent for wealthy collectors and museums seeking particular items" (Morris 2003: 32).

The history of the Umlauff firm ran parallel to that of Ward's, but as their enterprise was much more diverse, they appealed to a broader public, supplying customers who sought entertainment as well as clients interested in science. Although the firm gradually increased their focus on science supplies, they continued to provide items for the entertainment business:

Without giving up its origin in the entertainment industry, the Umlauff firm step by step grew to become more professional and began to supply an increasing number of educational institutions, university research centres, as well as natural history and ethnological museums. This dual orientation runs through the entire history of the Umlauff firm. (Lange 2006: 12)[3]

The Umlauff family became especially interested in apes, and gorillas in particular, as they developed their trade in mounted animals. Johannes Umlauff, the second oldest son of the firm's founder, Johann Friedrich Gustav Umlauff, was an eager gorilla hunter and a trained taxidermist. Johannes purchased apes brought to Hamburg harbour by ships sailing from Africa, and mounted them for sale, and he also travelled to Africa in order to acquire gorillas for taxidermy (Lange 2005: 183). The Gothenburg Natural History Museum has four photographs showing gorillas shot by Johannes Umlauff, but unfortunately additional information about time and place is poor.

CHAPTER 1

Fig. 8. Gorilla shot by Johannes Umlauff. Note the anthropomorphic position of the animal. The gorilla is tied to a frame and seems to be sitting. The animal's right foreleg has been arranged as if it is touching the left arm of the man standing in front of a group of young men with spears and machetes. The animal's body is slightly moved away from the white man and towards the Africans, linking the gorilla with the black man. The handwritten text on the backside of the photograph reads: "Der glückliche Schütze und der erlegte Gorilla" – "The happy hunter and the killed gorilla". Stamp: Johannes Umlauff, Naturalienhandlung und Lehrmittel, Eckernförderstrasse, Hamburg.

GNM_1984_1.

45

One of the numerous specimens produced by Ward's was the gorilla sold to the Gothenburg Museum in October 1906. Jägerskiöld's negotiations about price and quality had by this stage continued for more than a year. For a sum of £250, people in Gothenburg would have the opportunity to see a stuffed gorilla. The price included the skeleton (previously mentioned in the telegram), although the skeleton did not belong to the mounted gorilla, but rather to another gorilla killed in the French Congo. This does not seem to have bothered Jägerskiöld, or else it was overshadowed by the sheer excitement that a gorilla was almost at hand.

In 1906, the gorilla was in many respects a particularly significant species. One important aspect was its disquietingly human appearance, a pointer towards the so-called missing link between humans and apes. Another was the surprisingly delayed arrival of such a huge, new primate in museums and zoos. The western lowland gorilla was first described and presented as a new species by two Americans, the missionary and doctor Thomas S. Savage and the anatomist Jeffries Wyman, in 1847. This was the first modern description of any gorilla species. The first stuffed gorilla specimen was presented to the public by Paul Belloni Du Chaillu in 1861, just two years after Darwin's *On the Origin of Species* (Conniff 2011: 285–303). Fifteen years later the first live gorilla came to reside in Berliner Aquarium. The relatively late description of the gorilla greatly increased the demand for live and dead specimens at the time.

The mountain gorilla, a subspecies of the eastern gorilla, was only discovered in 1902, by the German, Oscar von Beringer, and described by Paul Matschie (also a German) who named it *Gorilla gorilla beringei*, later renamed *Gorilla beringei beringei* (Eckhart and Lanjouw 2008: 20). The animal's novelty had in 1920 fuelled Carl Akeley's motivation to organize his first expedition to study, photograph, and collect mountain gorillas, an expedition which took place the following year (Bodry-Sanders 1998: 176–177.). Compared with gorillas, elephants and lions had become commonplace. As Akeley stated:

To me the gorilla made a much more interesting quarry than lions, elephants, or any other African game, for the gorilla is still comparatively unknown. (Akeley 1923: 190)

There is little remaining documentation of the gorilla's passage from the field into the workshops of Rowland Ward. Ward's information about the gorilla is vague and scarce. They note that the animal came from Gabon, and that "we might mention that this particular Gorilla belongs to the redheaded race, which has received a specific name".[4] Neither do we know the name of the hunter, nor the date of the animal's death. Detailed information about the animals destined for sale could easily be lost in the flow of hides from the field to the taxidermy shops. An example of the sheer volume of animals being brought back from foreign hunting grounds is the thirty lions Ward's received for mounting in December 1912, quarry from Lord Delamere's hunting safari in Somaliland (Morris 2003: 39). The problem of the absence of evidence has been further compounded by the destruction of archived material. According to Rowland Ward's biographer, P.A. Morris, the firm's registers and job sheets have been lost in numerous workshop fires and during relocations to other sites (Morris 2003: 92). Morris mentions that Ward's "obtained specimens by purchase or barter, often from professional big game hunters" (Morris 2003: 38). One of these figures was the famous gorilla hunter, Fred G. Merfield (ibid.). No evidence, however, can be found to link the Gothenburg gorilla with Merfield. This means that the admittedly fragmented biography of the Ward gorilla begins in London.[5]

"The specimen is a good one, so far as Gorillas go"

In July 1904 Leonard Jägerskiöld had taken over as Head of the Zoological Collections. Although space was limited and the rooms crowded in the East India House, Jägerskiöld aimed from the very beginning at completing the collections with good examples of exotic beasts from remote corners of the world. His first enterprise as Head was to supply the collections with examples of three great apes: a gorilla, an orangutan, and a chimpanzee. As will be shown, obtaining the very best specimens for the Gothenburg Museum was not an easy task. While seeking new specimens, Jägerskiöld also initiated his ultimately successful campaign for a new building in which to house the zoological collections. In the 1909 Annual Report, Jägerskiöld warns the Museum Board that it would be impossible to fulfil the scientific ends of the collections, as well as the educational ones, if new and better conditions were not established within a reasonable time.[6]

CHAPTER 1

Fig. 9. The gorilla from Rowland Ward in the Gothenburg Natural History Museum.

GNM_5419_31.

The museum is today well-equipped with local specimens. The museum staff collected samples of Swedish fauna, Jägerskiöld accrued an important collection of marine specimens, while taxidermist David Sjölander went every year to Swedish Lapland to collect birds, nests, and eggs. Rare specimens from foreign countries, however, had to be bought from professional dealers abroad. Jägerskiöld had initially contacted Roland Ward Ltd. about the possibility of purchasing a mounted gorilla and an orangutan in June 1905. It appears that his letter was a response to a price list from Ward's. Jägerskiöld addresses the firm in his characteristically rough and direct way, unhampered by grammar:

Dear Sir.[7]

You offered in your last circular an Orang-utang £ 20 from Borneo. Is it an adult male? Can I get the skeleton also? I wish for my museum an adult male of *Simia Satyrus* and the skeleton from ape. I also wish an adult gorilla with its skeleton. Can you procure me these specimens? Your prices?

Yours truly

L.A. Jägerskiöld

Keeper of the Zoological Museum

The museum had, however, been a customer of Ward's prior to Jägerskiöld's letter. A letter from the London firm to the museum a few months before Jägerskiöld took up his new position as Head of the Zoological Collections indicates that the previous exchanges between Ward's and the museum were also based on barter. The exchange of specimens between natural history museums had long been common practice, often involving what to external observers may appear slightly bizarre items. For example, early in the nineteenth century, Paolo Savi, Director of Museo di Storia Naturale in Pisa, offered Arabian camels from the royal estate San Rossore, near Pisa, in order to acquire in return a platypus belonging to the Museum of Natural History in Geneva (Thorsen 2009). While the Bergen Museum in Norway had good access to whale skeletons that could be used for barter around 1900 (Brenna

2013), the Gothenburg Museum had elks. For an elk skeleton and skin, which "should be that of a fully adult male – nothing under 5 years of age, preferably of one nearer to 10 years of age", the museum could in 1904 exchange specimens from Rowland Ward Ltd. to a value of fifteen pounds.[8] In 1905 an adult male gorilla, mounted by Ward's, cost £150. If converted into elks according to the price estimated above, a mounted gorilla was worth ten full-grown male elk, including skeleton and skin.

We learn from Jägerskiöld's correspondence with Rowland Ward that the animals he wanted to procure for the Gothenburg collections should preferably be large males of the utmost quality. "First class" is a phrase frequently repeated in his correspondence with dealers. Ward's on the other hand tried to moderate their Swedish customer's expectations by both teaching him that old, male gorillas were not so good-looking, and that gorillas were part of a seller's market. The Rowland Ward company was astute in its ability to manage the media in order to get positive press (Morris 2003: 21). A similar shrewdness is displayed in their letters to Jägerskiöld. In January 1906 Ward's sent a letter with a photograph of a male gorilla, mounted in the same position as another they had prepared for Walter Rothschild's museum. The letter vouched for the mount's quality by referring to the typical appearance of old male gorillas: "Like all old males it is bare on chest and the head is slightly made up, but in all respects it is a very good specimen and an acquisition to any museum".[9] The price: £150. In February, Jägerskiöld briskly replied: "I ask you to let me know next time you get a good full sized Gorilla".[10] Ward found it necessary to remind Jägerskiöld that gorillas were not abundant: "No doubt we shall hear from you at the earliest possible moment and trust that we shall have a good specimen to send you. As you no doubt are aware, adult Gorillas are scarce, and for many years none passed through my hands."[11] Jägerskiöld has written with red pencil on this letter, "Intet svar" – no answer. In June, a new offer was sent to Gothenburg:

Dear Sir[12]

With further reference to our correspondence re Gorillas: We now have from the Gaboon (sic), a finely mounted specimen. The skin is in perfect condition, except – that – the hair is worn off the back head, as is the case, in all adult gorillas, which have passed through our hands. The skull has not been mounted in the animal, but we would send it with the specimen. We might mention that this particular Gorilla belongs to the redheaded race, which has received a specific name. Price (subject it being unsold when we hear) £ 250 (Two hundred and fifty pounds).

Thanking you
Yours faithfully
Rowland Ward
P.S. Photo can be sent.

Jägerskiöld found the price "somewhat elevated" compared to that of the gorilla offered to him in February. Nevertheless, he asks for the photograph and "the length head to tail and head to feet" – size was of importance.[13] In their answer, Ward's again emphasised that the specimen was representative of its species in that its fur was scarce in certain places of the body:

Dear Sir,[14]

We thank you for your letter. Gorillas "as skins" are being sold on the Continent for £400 to £500 each, so our price is certainly not elevated. The specimen at £150 pounds (sic) has already been sold. We send herewith a photograph of a fine male Gorilla, which we have for sale.

The specimen is a good one, so far as Gorillas go, that is to say, it is as perfect as the majority. Some hair has been supplied and added to the shoulders and top of head where they invariably "slip". The length of the animal from the crest of the head to the anus is 38 inches, the length of the lower limbs from the crutch to the longest point 28 inches, the reach of the arm from the base of the neck to the highest point where it is hanging 34 inches, girth of the chest 55 inches, girth of the upper arm 19 ½ inches.

The skull is separate. Price, subject to being unsold, is £250 (two hundred and fifty pounds).

Thanking you
Yours faithfully
Rowland Ward Ltd.

Apparently Ward's played with open cards by admitting that the mount had been embellished. Fur had been added on top of the head and on the shoulders. But, if adult gorilla males normally lacked fur on their shoulders, why had hair from another animal been added? Certainly this made the specimen less authentic, but more aesthetically pleasing to the public eye. What Ward calls a "slip" was a physical imprint of the male western gorilla's habits. Paul Belloni Du Chaillou, who claimed to be the first white man in modern times to see and hunt the gorilla,[15]

describes the habits of the great apes of Africa in some detail in his *Explorations and Adventures in Equatorial Africa*:

Only the young gorillas sleep on (sic) trees, for protection from wild beasts. I have myself come upon fresh traces of a gorilla's bed on several occasions, and could see that the male had seated himself with his back against a tree-trunk. In fact, on the back of the male gorilla there is generally a patch on which the hair is worn thin from this position, while the nest-building *Trolodytes calvus*, or bald-headed *nshiego*, which constantly sleeps under its leafy shelter on a tree-branch, has this bare place at its side, and in quite a different manner. (Du Chaillou 1861: 395)

Western gorillas are known to make nests from plants in which to sleep, fashioning both a night nest and a day nest. Most probably, the scarcity of hair on the back is due to rubbing against tree trunks. In other words, important information inscribed in the live gorilla body had been erased from the mounted specimen. Moreover, what did it actually mean for a specimen to be a 'good gorilla", in the sense that it is considered consistent with the majority of gorillas – was the animal in question, therefore, a mediocre gorilla?

Jägerskiöld remained strategically silent for a couple of months, until Ward wired that the museum could purchase the mounted gorilla with its skull separate, plus a mounted skeleton from another gorilla, for the price of £250.[16] Jägerskiöld promptly wired back: "Please keep offer firm one week while collecting necessary amount."[17]

The price of the mounted gorilla greatly exceeded what could be acquired with the museum's annual budget; the sum would have to be found outside the museum. As many as fifteen of Gothenburg's most prominent citizens contributed to the acquisition, some donating as much as 1,000 Swedish crowns, such as the Director of the shipyard Götaverken James Keiller senior (who also donated the money needed to buy and mount an orangutan hide) and consul C. Lyon. Other donations were on a scale from 1,000 down to 100 Swedish crowns. In 1906 the museum received private contributions amounting to 6,350 Swedish crowns, 3,300 of which covered the transaction with Rowland Ward. In total the amount spent on acquisitions in 1906 ran to 9,524.79 Swedish crowns. This indicates that the Zoological Department was heavily dependent upon private donations in order to expand their collections, and that there were wealthy citizens willing to support the Gothenburg Museum's efforts to collect natural history specimens.

At the end of 1906 the museum owned 587 foreign mammals, and 282 Swedish mammals.[18] Jägerskiöld, who had used a third of the total sum paid for new objects in 1906 to buy the gorilla and the gorilla skeleton, wrote in the annual report that the purchase "compared to what approximately at the same time and in other places was paid for the same species, must be considered extremely cheap."[19] Nearly forty years later Jägerskiöld described the procurement of the gorilla as "my first more extensive beggary", a "gorgeous piece, offered to me for £250 from a big, famous London firm. I telegraphed: 'I am collecting money.' Totally a sum of 5,000 Swedish crowns was required."[20] This figure, a slight exaggeration, may be due to the years that had passed, or maybe Jägerskiöld had converted the sum to the value of the crown in 1942/43 in order to better underscore the animal's value.

In the final paragraphs of the 1907 annual report, Jägerskiöld clearly states that if his departmental expenses seem high, it is because the Zoological Department is the only one in the museum that "manufactures museum objects on a broader scale", referring to ongoing taxidermy work.[21] Apparently, in his understanding of a museum object, he distinguished between mounted specimens in zoological displays; illustrative, pedagogical objects, made to educate and inform the public about natural history, and the historically authentic items presented in the other museum departments. The purchases and the production of museum objects were part of Jägerskiöld's ambition to transform both the display of the zoological collections and the inner organization of his department to serve a public audience, teachers, and scientists. The purchase of the gorilla was part of this modernization. At the end of 1907, the museum held examples of three of the great apes: the gorilla purchased from Rowland Ward Ltd., an orangutan, also bought in 1906 from Georg Albert Girtanner, and a chimpanzee hide from Cameroon, mounted by Hilmer Skoog.

"Kindly tell your staff not to handle the stuffed Gorilla by its limbs", this underlined warning was sent to Jägerskiöld from London before the specimen was shipped to Gothenburg.[22] Hilmer Skoog was himself an experienced taxidermist and knew how to handle the gorilla when he unpacked the body and put it on display on 17 and 18, October, 1906.[23] The purchase of the gorilla proved to be a good investment, as the mount quickly attracted the Gothenburg public. During the two

hours the museum was open on the first Sunday, the so-called 'vernis-
sage of the gorilla', the specimen was seen by more than 2000 people
(Jägerskiöld 1943: 347).

"Wir können nur erstklassiges brauchen"

Jägerskiöld's extensive correspondence with Rowland Ward had re-
volved not only around gorillas, but also around the possibility of ac-
quiring a good specimen of the male orangutan. Ward had supplied
Jägerskiöld with a photograph of a newly mounted orangutan specimen, hanging by both arms to show the position in which the animal
would be fixed.[24] Again, Jägerskiöld appears to have been a sly dealer on
behalf of his museum and its meagre budget. Parallel to his correspondence with Ward's runs a similar exchange with Swiss Georg Albert
Girtanner (1839–1907) in St. Gallen about the possibility of purchasing
a good gorilla and an orangutan. Jägerskiöld's involvement in a double
negotiation may explain the impatience he expressed in some of his
letters to Ward's, where he asks about the possibility of purchasing an
orangutan from London:

When you offers (sic) me some animal I would be glad to get at once the price
of skin and of stuffing. Size and (unreadable word). This will save a good deal of
writing.[25]

It took nearly two years, thirteen letters, and three telegrams before
the transaction between Jägerskiöld and Ward's was settled in October

1906. While Jägerskiöld was very satisfied with the gorilla, the orang-utan came as something of a disappointment.

In June 1906, Jägerskiöld's order for an orangutan went to Girtanner, and not to the business minded Ward's. Girtanner was a medical doctor by training and practice, but also a talented and internationally acknowledged amateur ornithologist. As was the case with many of his contemporary naturalists, Girtanner's efforts to save indigenous species from extinction did not interfere with his dealing in dead animals. Girtanner was a member of the natural history museum board in St. Gallen, and through his contacts with museums in Switzerland and abroad, with the zoos in Vienna (Schönbrunn), Paris (Jardin des Plantes), and in London (London Zoo), as well as with scientists, hunters, and dealers in naturalia, he procured specimens for the museum in his home city (Brassel 1908). Girtanner also sold specimens to foreign museums, and the exchange of letters between him and Jägerskiöld indicates that this was not the first time the two gentlemen had done business together.

In a letter dated 7 May, 1907, Girtanner expresses his thanks for having received a cheque for 600 francs, or 435 Swedish crowns, from Jägerskiöld for "den unglückbelasteten Orang", the unfortunate orangutan.[26] Even if Jägerskiöld had worked very hard to acquire a good orangutan specimen for the collection, the result was for various reasons far from what he had expected. Jägerskiöld had bought the skin from Girtanner in June 1906.[27] According to the General Register, the Swiss doctor had obtained the skin from somewhere in "America", and the animal was supposed to have been killed in its habitat.[28] Girtanner also arranged for the mounting to be executed by a German taxidermist, Herr Bauzer, living in Oehringen. One and a half years after the purchase of the skin, the orangutan had finally arrived at the museum. On 28 January, 1908, Skoog unpacked the specimen and trimmed it.[29] The long-awaited orangutan that Jägerskiold had worked so hard to obtain was a major disappointment. Two days later, Jägerskiöld ends a letter to Girtanner with the sentence: "Ich bedaure dass ich diesmal ein so unangenehmes Brief habe schreiben mussen aber wir können nur erstklassiges brauchen."[30]

What was wrong with the mounted orangutan? Jägerskiöld found its position to be good, but the face was if not bad, so definitely not good. Even worse was the coat, which he had expected to be of a very high quality. Jägerskiöld had been promised "eine sehr schöne Haut von Orang utan (sic)"– a very beautiful orangutan hide. Instead, a closer inspection of the hide revealed that pieces of fur had been plastered onto the body, especially across the head, and in a short note to the taxidermist, Bauzer, Jägerskiöld postulated that "beinahe d. halbe Tier mit falsche Haar besetzt ist (sic)" – approximately half of the animal is plastered with false fur. Previously, Girtanner had only delivered good specimens, so what was the reason for sending him this one? Perhaps Girtanner had not examined the specimen himself?[31] Girtanner answered immediately. In an extended letter he expressed his deep sympathy with Jägerskiöld. Girtanner himself had not been involved with the expedition of the specimen to Gothenburg, and had thus not seen the finished mount. On the other hand, he had always allowed only first class mounts, and the taxidermist Bauzer, who had made the

Fig. 10. Head of "der unglückbelasteten Orang-outang".

orangutan, had always delivered good specimens in the past. Never-theless, the poorly mounted ape brought Bauzer's commissioned work for Girtanner to a close.[32] The unlucky taxidermist addressed himself to Jägerskiöld as well as to Girtanner, with the extremely odd advice to treat the fur and skin with turpentine and oil.[33] Girtanner himself died a few months later.

CHAPTER 1

"Where do you take eyes from?"

The three great apes in the Mammal Hall at the Gothenburg Natural History Museum clearly demonstrate that exotic animals were transformed into scientific objects through geographically complex trajectories. The gorilla had been shot in Gabon and mounted in London, the orangutan had been killed somewhere in Indonesia and its hide passed through "America" before being sent to Girtanner in St. Gallen and then to Bauzer in Oehringen.

The third of the great apes, a chimpanzee, was naturalized by the museum's own taxidermist, Hilmer Skoog. The hide and skeleton had been acquired by G. Linnell in 1904 and given to the museum. The full grown male had been killed near the village of Debundscha, Cameroon.[34] Skoog completed the job in 1906–1907 to Jägerskiöld's full satisfaction: "The mount of the male chimpanzee from Gabun (sic) already purchased last year has been finished this year by taxidermist Skoog, and is a substantial proof of his skill and its accomplishment is just as good as the gorilla mounted by the world famous firm Ward, bought in 1906".[35] Here Jägerskiöld mixes the chimp's provenance with that of the gorilla.

We can follow Skoog's work on the chimpanzee through his journal entries. On 30 November, 1906 he "drew chimpanzee for stuffing". The mask for the chimp he modelled from sawdust on 8 March, and on 27 March he had completed and set up the body. From 11–13 April he fastened the skin on the manikin: "Sewed the hide of the chimpanzee".[36] It is clear that Skoog used the dermoplastic method for his taxidermy, because in 1916 he explained to the press why modern

Fig. 11. The chimpanzee mounted by Hilmer Skoog.

PHOTO: ANDERS LARSSON, GMN.

taxidermy was an art: They no longer 'stuff' animals, but their works consist in various types of skin or dermoplastic modelling.[37]

After the transaction with Ward had been settled and the arrears for the gorilla had been paid, Jägerskiöld made an effort to wring from the famous taxidermist some secrets of his craft. Jägerskiöld's concern was with glass eyes for chimpanzees, and artificial gorilla penises. Skoog was to finish the chimpanzee, and was looking for good eyes.

Badly chosen eyes inevitably make "bad taxidermy" (Morris 2010: 169). The gorilla's penis is today subject of conversation among the museum staff because it is supposed to be omitted by the taxidermist, but in this case, it is not. The penis is tiny compared to the size of the animal, but anatomically correct. Jägerskiöld was also curious about the gorilla's "viril (sic) member", not about its size, but about how it had been made:

Dear Sir.

I wonder if you can procure me a pair of eyes to an adult male chimpanze (sic). Where do you take eyes from? I have taken German ones but I am not quite satisfied with them. I am very curious to know the composition of the mass you use in modelling for instance the penis of the last sent male gorilla when the viril (sic) member was modelled quite artificially. When you get a good stuffed specimen of the *Rhinopithecus roxellanae* (sic) A M Edv. please let me know and your price for it as also sex and size.

Yours very truly

L.A. Jägerskiöld[38]

Jägerskiöld is entering forbidden territory here. Rowland Ward Ltd. was not a firm that divulged their secrets to competitors in the trade. It was, for instance, strictly forbidden to circulate photographs showing the mounting process; only pictures of finished specimens were sent to the press (Morris 2003: 6). Ward's answer, however, is utterly polite. In an exceptional gesture, Jägerskiöld is given the eyes for free: "In reply to your letter of the 3rd inst., we do not in the usual way supply eyes, as nearly all our work of this kind is special. However, by sample post we send you a pair suitable for the Chimpanzee." The composition they use is, however, their own special one, read: not to be told to anybody else. As for the gorilla's penis, it was, as far as they could remember, "made up of a real piece of skin". [39]

Today the gorilla from Ward's, the orangutan from Girtanner, and the chimpanzee mounted by Skoog are displayed together in a glass case in the Mammal Hall, just as they had been presented when the zoological collections still were located in the East India House. They were later joined by a poor little fellow, an infant chimpanzee that died in Gothenburg in 1892. The sight of this animal is extremely sad. Many of the captive chimpanzees and orangutans who were taken to Europe during the nineteenth century died of tuberculosis and other illnesses due to malnutrition and the cold climate, and few survived more than a few months or a year (Ritvo 1987: 31).

Photography and Gorilla Positions

In 1909, Umlauff wished their customers a happy new year by sending them a card adorned with a mounted gorilla photographed from three angles. At the end of the nineteenth century, photography was adopted as a mimetic technique by different but intertwined spheres such as art, display, taxidermy and trophy making - and for the presentation of commercial commodities. Cameras were soon used on geographical explorations, by artists making preliminary studies, and by taxidermists documenting animal positions and details in landscapes.[40] Many of these landscape details were later reconstructed in the museum workshop

Fig. 12. New year's greetings card from Umlauff 1909.
Commercial for a mounted male gorilla.

GNM_232_1.

and the diorama; in many ways photography became a substitute for drawing.[41] Michelle Henning points at the confluence of photography and taxidermy: both are techniques that freeze animal motion. The photograph was also the basis of the diorama as it "painstakingly reconstructs the traces and details that are indifferently recorded by machines such as the camera" (Henning 2006: 52). As Constance Areson Clark states: "In the late nineteenth and early twentieth century many scientists sought methods of illustration, such as photography, which would demonstrate their objectivity by removing the scientist as an individual from the cognitive process of technical illustration" (Clark 2008: 28).

Jägerskiöld used photographs in his zoological displays in a different way than we tend to see them in today's museum displays, where enlarged pictures recreate a milieu or backdrop for historical items. In anticipation of the arrival of the animal body, Jägerskiöld displayed a picture of the species he wanted to display. Photographs made space for animal materiality. Jägerskiöld wrote his first report for the Zoological Department early in January 1905. He had then been Head of Department for only six months. The report reflects his strong motivation to improve the conditions of the collections and the exhibitions. We read that Jägerskiöld had procured enlarged photographs to better explain the animals' natural behaviour to the public, especially that of Swedish birds, which were a prominent research subject during the years he was in charge of the department. Of special interest when discussing the various stages of obtaining exotic specimens, is the fact that, apart from the bird photographs, Jägerskiöld had in 1904 also purchased large images of foreign animals, species he knew would be physically out of reach for a long time. These included elephant and rhinoceros, specimens that were not obtained for the museum until 1948 with Sjölander's expedition to Portuguese West Africa.[42]

Jägerskiöld's procurement of the photographs was the first step to bringing "missing" big African fauna into the museum's exhibition. While the photograph surely represented the actual animal, it simultaneously created an expectation of replacement and a vacancy waiting to be filled with "the real thing." Taxidermist David Sjölander's long nourished dream to collect and mount an African elephant certainly reflected a personal ambition, but it also involved replacing a photograph with the materiality and sheer volume of the mounted beast; it was both idiosyncratic and rooted in the history and agenda of the institution (see Chapter 4).

The photographs in the zoological displays expressed Jägerskiöld's visions for his museum, photographs sent from Ward's offered him opportunities to realize these visions. In the letters from Ward to Jägerskiöld there are frequent references to photographs of mounted animals for sale, photographs that the firm had either sent to Gothenburg or was prompted to send. Some of these pictures, of a gorilla, a chimpanzee, and an orangutan, are held in the museum archives.[43] Rowland Ward was himself a devoted photographer, and for his documentation of zoo animals, he invented a portable camera (Morris 2003: 16). As P.A. Morris observes, "Rowland Ward was a very effective publicist" (2003: 30). He promoted his business by advertising in the newspapers and sending price lists and individual handwritten letters to natural history museums around the world. He also furnished the newspapers with photographs of a noteworthy animal whenever a mount had been completed (Morris 2003: 21).

Ward's correspondence with Jägerskiöld reveals that sending photographs of specimens for sale to potential customers was another efficient way of advertising. Instead of sending stuffed gorillas across the ocean to and from Sweden, photographs of them could function as trade samples. The photographs were not only trade samples, however; they could also work to establish confidence about the specimen's authenticity and the quality of the taxidermy, and thus certify the zoological value and public appeal of the specimens themselves.

The photographs constitute important evidence of how "a good gorilla" specimen was defined by taxidermists around 1900, suggesting differing interpretations of the animal. Nigel Rothfels has argued that the classic gorilla position that "echoed in museums around the world for a century" was that shown in the frontispiece of Du Chaillu's book *Explorations and Adventures in Equatorial Africa*, published in 1861 (Rothfels 2002: 3). A gorilla with bared teeth is shown standing upright while holding on to a tree branch with one arm, and the angry animal seems to turn towards an enemy. 1861 was also the year in which Du Chaillu presented the first stuffed gorilla at the podium of London's Metropolitan Tabernacle, "placed in the attitude of speaking, with its arms outstretched and its hands grasping the front rail of the platform", as described by the London daily *The Times* (Conniff 2011: 285).

The German historian Britta Lange has thoroughly investigated the iconography of one famous gorilla preserved by the brothers Johannes and Wilhelm Umlauff, called the giant gorilla, in German "Riesen-Gorilla", a lowland gorilla claimed to be of dimensions previously unseen. In the summer of 1900, Johannes Umlauff bought the skeleton and hide of an enormous gorilla from a Mr. Paschen. Paschen had shot the animal in Cameroon, and had returned to the port of Hamburg with his trophy (Lange 2006: 88). The Umlauff brothers chose to present the gorilla in a position characterised by Lange as "Trophäen-Ikonographie", trophy iconography (Lange 2006: 108). By choosing to arrange the gorilla into a aggressive position, the Umlauff brothers reintroduced "das alte Motiv der Bestie", the old motif of the beast (Lange 2006: 96). Indeed, the position of the giant gorilla mimics the Du Chaillu illustration of forty years prior. The Umlauff gorilla is erect, its mouth wide open, its right leg resting upon a stump, while its left hand is clenched. Lange argues that this gorilla confronts an imaginary hunter, meaning that the audience experiences the animal as the hunter did seconds before its body was struck by a lethal bullet (Lange 2006: 97). The giant gorilla was put on display in "Umlauff's Weltmuseum", the museum of the world, and in January 1901 it was presented at the *Jagdaufstellung* in Berlin before appearing in several other German cities. In August 1901 the gorilla was displayed at the V. Internationalen Zoologen-Congress in Berlin. One of the participants was Walter Rothschild, who bought the mounted specimen and the skeleton for the then astronomical sum of £1,000, four times the price Jägerskiöld would accept to pay for the Ward gorilla and a skeleton five years later (Lange 2005: 191 and footnote 25, Lange 2006: 5).[44] Rothschild had a particular fondness for gigantic specimens, and "(h)is museum displayed the largest gorilla specimen known" (Conniff 2011: 330). This purchase moved the "Riesen-gorilla" from an atmosphere of entertainment in Umlauff's "Weltmuseum" to the scientific displays in the Tring Museum. Not only was the reconstructed body of an extraordinarily large gorilla displayed in a new setting, but more importantly, a specific representation of the gorilla was introduced to an influential milieu: the gorilla as a ferocious beast was reintroduced to the museum. With the transfer from Hamburg to Tring, the conception of the gorilla as a monstrous and ferocious beast would now be presented to audiences in a scientific context. The "Riesen-gorilla" is still to be seen in the Natural History Museum at Tring.

Fig. 13. Dead gorilla placed on a box. Killed 19 October, 1914, in Mbassaland.

PHOTO: UMLAUFF. GNM_1981_1.

"Umlauff's Weltmuseum" had opened at the end of 1889, displaying groups of wax figures from exotic countries, ethnographic items and preserved animals. The museum was, Lange maintains, a strategic way to generate publicity for the rather anonymous firm J.F.G. Umlauff, and it became the firm's showcase. During the 1890s, Umlauff succeeded in selling model figures of exotic people to German ethnographic museums and to standardise the production (Lange 2006: 57). The firm continued to play a role as a trading company: "This favoured systematizing the offer of figures and its serial production for exhibition purposes, even without a museum of its own" (Lange 2006: 84).[45]

Which were the gorilla positions preferred by Umlauff? There are twelve photographs of gorillas from Umlauff in the archives of the Gothenburg Natural History Museum. The photos depict either the dead animal in some African village, or mounted gorillas in Umlauff's museum. It is important to note that none of the mounted gorillas seen in the photographs are snarling like the "Riesen-Gorilla". However, it is possible to associate the positions given to the newly killed animals with the positions given to the mounted gorillas in Hamburg. Three of the dead apes are arranged in seated positions, one of them having been tied upright to a frame. The arrangements of the dead bodies in the field anticipate the positions given to mounted gorillas in the museum.

Lange argues that Umlauff's gorillas speak not to the animal as such, but to the relation between animal and human: "In this context even pictures of animals were pictures of humans" (Lange 2006: 99).[46] This point goes for the gorillas that feature on the Umlauff photographs examined here. The Ward gorilla in Gothenburg is hanging by one

Fig. 14. Dead gorilla placed on a garden chair.

PHOTO: UMLAUFF. GNM_1986_1.

arm. Umlauff delivered gorillas in very different positions. Instead of focusing on the gorilla's ability to move between treetops, an ability that separates it from man, the taxidermists in Hamburg chose to highlight the animal's likeness to man – the black man.

Umlauff delivered gorillas standing or seemingly walking upright. Some are sitting on a stump like the gorilla cadaver photographed in Africa. On the New Year's card from 1909 an erect 'walking gorilla' supporting itself with a club in its left hand, wishes customers "Die besten Wünsche zum Neuen Jahre". The gorilla is presented face on, from behind, and from the side – and is obviously a commercial object. On the card customers were informed of the quality of the gorilla's fur on each part of its body, its height, and told that it had been killed in the border districts between northern Cameroon and French Congo. The composition resembles a picture of a distant, human forefather. Two other pictures from the Weltmuseum portray a hall filled with glass cases in which gorillas are standing or sitting on stumps, as if performing various tasks. Gorilla skeletons are displayed in between the cases. These gorillas resemble human figures as usually portrayed in ethnographic exhibitions. The connection between so-called "primitive peoples" and apes was in this case explicit. The gorillas dominate the hall, but a closer look reveals that models of black people mingle with those of the apes; a young boy sitting, and a warrior with a spear (see Fig. 15).

Umlauff mounted gorillas in the Du Chaillu tradition of the ferocious beast, and also, as several of the photographs indicate, arranged gorillas in less threatening positions. Lange states that Wilhelm Umlauff, who composed the "Riesen-Gorilla", worked after photographs

Fig. 15. Mounted great apes for sale in the Umlauff Weltmuseum. Note the composition of a chimpanzee family in the foreground. In the background two models of black men, one sitting and the other standing with a spear.

PHOTO: UMLAUF. GNM_1979_1.

Fig. 16 Anthropomorphic gorilla couple.

PHOTO: UMLAUFF. GNM_1982_1.

Fig. 17. Mounted gorilla. Johannes Umlauff sent Jägerskiöld the photograph in a letter dated 14 June, 1929, in which he informs that the newly made "Gorilla-Männchen" – the little "gorilla man" – is for sale at a cost of 5,000 marks. The example is claimed to be "erstklassig"– first class. [47]

PHOTO: UMLAUFF. GNM_1983_1.

sent to him by the gorilla expert Paul Matschie, then head of the mammal department at Königliches Berliner Naturkundemuseum (Lange 2005: 192). Despite this connection to scientific expertise, Umlauff's gorilla taxidermy represented an anthropomorphic ape placed on the ground, erect or sitting.

Rowland Ward offered a version of the gorilla very different to that of Umlauff. Ward was the pre-eminent British authority on handling big game for trophies, although he had never been to Africa or Asia himself (Morris 2003:16). What was considered a 'good' posture for gorillas at Ward's? As revealed by the correspondence in Gothenburg Natural History Museum and the photographs Ward's sent the museum in 1905 and 1906, Ward had specialized in hanging gorillas, orangutans, and chimpanzees. In Ward's answer to Jägerskiöld's letter of 26 June, 1905, they describe the posture of an orangutan as "[w]e should prepare doing the orang hanging".[48]

Hanging gorillas are also mentioned in letters Jägerskiöld later received from Ward. The museum has four photographs from Ward's: two of gorillas, one of a chimpanzee and one of an orangutan. All of them are hanging by one or both arms from a piece of wood fastened to two strings, like a trapeze. In their allusions to circus artists, these postures transport the apes from their African and Asian habitats, and lives of relative liberty, to the world of human entertainment and submission. Rowland Ward had never seen gorillas in their natural habitats. But on his frequent visits to London Zoo he studied the great apes in their artificial surroundings, and reproduced them as such in his taxidermy shop. Probably this hanging position was chosen for

Fig. 18. The Ward gorilla in the Gothenburg Natural History Museum.
Note the peaceful expression, and the visible seam on the breast.

specimens destined for museums, while trophy animals must have been made according to the owner's wishes. The Ward gorilla, hanging by its arm, is impressive, but its awe-inspiring strength is countered by its peaceful countenance.

Gorillas were displayed as mounted specimens and as skeletons. In natural history museums, as well in influential zoology books, the gorilla skeleton has often been shown alongside and in comparison to a human skeleton. To achieve the most instructive presentation of the similarities and dissimilarities between the two species, the gorilla skeleton was made to stand erect to match its close human relative. The (sometimes) lower and coarse-limbed gorilla skeleton and the delicately built human skeleton matched each other and became a standardised display genre, a kind of natural history's Laurel and Hardy.

In Gothenburg Natural History Museum a case called "Animal versus Man" contained a gorilla skeleton and a skeleton of a *Homo sapiens*.[49] The German firm Dr. Schlüter & Mass, late nineteenth and early twentieth century dealers in natural history specimens, offered gorilla and human skeletons arranged together.[50] The skeleton couple also features in Alfred Brehm's famous reference work *Illustrierte Tierleben* (1863–1896), later known as *Brehms Tierleben,* written for homes and schools and published in English and Swedish. Thus people became accustomed to seeing and thinking of the gorilla alongside man, erect like him, but inferior to man in height, delicacy and the proportion of the brain, a proletarian compared to the noble looking man (Löfgren 1985: 197).

The Roar of the Gorilla

Du Chaillu's horrifying and dramatic descriptions of his meetings with gorillas nourished the popular understanding of the species for decades. Claiming to be "the first white man" to eyewitness the animal, Du Chaillu's authority to describe the great ape was further bolstered by the earliest scientists, such as Richard Owen in his *Memoir on the Gorilla* of 1865. On the one hand, Du Chaillu claimed that many reports of the gorilla were untrue: the gorilla did not stalk and kill human prey, nor did it kidnap women from the villages. On the other hand, he strongly emphasised the animal's monstrosity: "(N)o description can exceed the horror of its appearance, the ferocity of its attacks, or the impish malignity of its nature" (Du Chaillu 1861: 394). Du Chaillu depicts the horror of the gorilla as being beyond imagination:

Then the male, sitting for a moment with a savage frown on his face, slowly rises to his feet, and, looking with glowing and malignant eyes at the intruders, begins to beat his breast, and, lifting up his round head, utters his frightful roar. This begins with several sharp barks, like an enraged or mad dog, whereupon ensues a long, deeply guttural rolling roar, continued for over a minute, and which, doubled and multiplied by the resounding echoes of the forest, fills the hunter's ears like the deep rolling thunder of an approaching storm. I have reason to believe that I have heard this roar at a distance of three miles. The horror of the animal's appearance at this time is beyond description. It seems as monstrous as a nightmare dream – so impossible a piece of hideousness that, were it not for the danger of its savage approach, the hunter might fancy himself in some ugly dream. (Du Chaillu 1861: 396– 397)

How long did it take before "the roar of the gorilla" abated? And to return to the Ward gorilla in Gothenburg, what did early twentieth-century visitors think of the great ape? The sources that might help answer these questions are poor. Surely they perceived the gorilla in ways very different to those of today's public. As Ritvo has observed: "Anglophones tend to feel closer to gorillas and chimpanzees now than they did in the late nineteenth century", explaining this emotional shift in terms of the knowledge we now have of the great apes' biology and behaviour (Ritvo 2010: 11). Today we stress the similarities between us and them as something positive. One hundred years ago, however, these similarities provoked images of the great apes as travesties of the human (Clark 2008: 3).

In her study of the J.F.G. Umlauff firm, Britta Lange demonstrates that there were two competitive ways of perceiving the gorilla around 1900; the ferocious beast as figured by Du Chaillu, and a conception of the gorilla as friendly and docile. This secondary counter-narrative was nourished by the presence of live gorillas in European cities, the first of which arrived in Berlin in 1876: "Since it had been possible from 1876 on to keep the gorilla Mpungu alive for more than a year in the Berlin Aquarium, emphasis was put on the animal's peaceful and human-like appearance" (Lange 2006: 96).[51] After spending less than two years in the Berliner Aquarium, Mpungu shared the destiny of so many zoo-bound great apes, and died of tuberculosis.[52] Mpungu, before he arrived in Berlin, had been raised by humans. In January 1881, *The Popular Science Monthly* published a short piece on "Studies of Young Apes" in which Mpungu figures. The text emphasises how easily the gorilla adapted to humans, and noted his friendliness:

Mpungu gave a contradiction to the reports of the fierce and untamable character of the gorilla, for he soon became accustomed to the persons around him, showing a real dependence upon them, and confidence in them, and was allowed to run about with no more care than a child.[53]

Despite Mpungu's less than two years in Berlin, the public learned from watching him that gorillas were peaceful creatures. Gorillas in captivity worked to undo many of the phobic assumptions held by the public since Du Chaillu's portrayal of the animal. Other famous zoo gorillas, such as Alfred the Gorilla, had similar effects on public conceptions of gorillas. Alfred was born in the Congo. He was orphaned as a baby, and had been nursed by a local woman. The Bristol Zoo bought him in 1930. When Alfred died in 1948, he was the longest living gorilla in captivity. His skin was mounted by Ward's (Paddon 2011: 137). Hannah Paddon describes Alfred as a mascot animal, and like many other mascot animals, e.g. Monjet (discussed in Chapter 2), he has been preserved and is today showcased in the Bristol City Museum and Art Gallery (Paddon 2011: 136). His immense popularity "helped to curb, and ultimately vanquish, the cultural and societal misconceptions of the gorilla painted by the likes of Paul Du Chaillu, who had proclaimed the animal to be a "hellish dream creature", and Reverend John Leighton Wilson, who had described the species as "one of the most frightful animals in the world" (Paddon 2011: 136).

Carl Akeley, who had met gorillas in captivity, considered the gorilla to be "normally a perfectly amiable and decent creature" who attacks man "because he is being attacked or thinks he is being attacked" (Akeley 1923: 196). The historical narratives of zoo gorillas show that these animals, during their captivity, paved the way for a more balanced and sober understanding of the species.

The Gothenburg Museum bought the mounted gorilla in 1906. He is not snarling at the viewer, nor is he standing erect, nor is he banging his chest. The docile looking beast had been hanging in the glass case for almost twenty years when an anonymous journalist from the newspaper *Göteborgs Aftonblad* visited the museum and wrote a piece titled "Tjocka släkten" or "Closely related":

The gorilla uses tree branches to protect himself, but his most frightful weapon is simply his arm. A single blow of a gorilla fist can make its adversary fall stone dead to the ground. The enraged gorilla male is a horrid revelation, there he comes through the forest walking erect on his hindquarters, banging his fists on his barrel-like chest and letting out grotesque roars from his open throat, the baring canines as large as those of the carnivores. Hands and teeth are also the weapons of the gorilla against its enemies in the woods.[54]

At the time when the Ward gorilla was mounted, scientists in the US and in Germany mostly agreed about evolutionism (Clark 2008: 42, Lange 2005: 187). The text quoted above was written in the 1920s when the question of man's descent whether it be "God–or gorilla" was again fervently debated. Scientists' pros and cons of evolution echoed in

American popular culture, in images as well as in jazz music (Clark 2008). Maybe it is far-fetched, but it is tempting, to read the journalist's utterance as typical of the multilayered popular understanding of the gorilla. The journalist admits that the gorilla is man's close relative. But it is a brutish relative; in his description of this docile representation of the gorilla the journalist evoked the image of the roaring monster, thus echoing both the old Du Chaillu's rhetoric and the many contemporary images of an erect, walking aggressive ape, featuring in the Tarzan movies of the 1920s, and culminating with King-Kong in 1933 (Clark 2008: 11–12, Lange 2006: 123–127).

Commodity Animals

What is to be learnt from the Ward gorilla? Perhaps most notably, it exemplifies how, at the turn of the twentieth century, museum specimens made for sale passed through so many different hands, moved through so many different spaces, and underwent so many morphological transformations, that it is impossible to re-render the animal's individual traits or "authenticity". The Ward gorilla, as well as many of the gorillas made by Umlauff, also demonstrate that around 1900 taxidermy firms developed various representations of gorillas destined for the market, offering the ferocious Du Chaillu position and competing portrayals.

More than one hundred years have passed since the Ward gorilla and the badly mounted orangutan arrived at the Gothenburg Museum. They are still on display. A closer look at the orangutan reveals that fur has been added to the back of the head and the shoulders, as

Jägerskiöld had complained to Girtanner. Similar embellishments are not discernible on the gorilla, although we have learned that this specimen was indeed altered to improve its aesthetic appeal. Is the gorilla mount good handicraft, in the sense that the Ward's taxidermist did a better job than Herr Bauzer? A taxidermist's skill can be judged by looking to the exposed regions of a body where seams cannot be hidden by fur. A gorilla's breast lacks fur; on the Ward gorilla, the seam on the chest, and the seam that runs from the neck to the base of the torso, are both very visible. The Ward gorilla and the orangutan also testify to the fact that the skin of an animal was restored with a focus on hiding scars and marks accrued during its natural life, in addition to those sustained during its slaughter. Thus the taxidermist aimed at enhancing animal bodies from normality to perfection, to a standard that customers such as Jägerskiöld were seeking: the "first class" specimen.

The posture of the gorilla's body is stiff and lacks plasticity. Knowing that it was mounted into a position that the taxidermists had repeated several times, the gorilla can be seen as a standardised handycrafted commodity. The Ward gorilla is a 'glass case' specimen, designed to fit within the natural history museum. Its position was not based on the taxidermist's personal observation of gorillas in their natural habitats. As explained above, Rowland Ward had never seen a gorilla in the wild. In this sense, his approach was significantly different to that of Carl Akeley. Akeley modelled his famous mountain gorilla diorama in the American Museum of Natural History with material he collected in the field. He carefully selected each individual animal according to the intended composition of his diorama. These animals

Fig. 19. *Orangutan bodies in "Lager Kühlhaus".*

PHOTO: UMLAUFF, 1914. GNM_1994_1.

were arranged for singular, unique displays in which each specimen comprised part of a greater unity.

When I was examining the Umlauff photographs in Gothenburg Natural History Museum's archive, three pictures expressing a motif I had never seen before, turned up. The pictures show various ape and monkey bodies hanging by metal strings fastened around their necks. The bodies are complete. The orangutans' limbs are crossed, as if they are trying to protect their fragile bodies, even in death. Some of them look as if they are asleep, while others grit their teeth. The species is represented by adults and infants. In two of the photographs, every ape has a numbered label fastened to the body; the numbers run from 429–437. Why were these apes photographed? What was the value of these corpses? How did the museum obtain the photographs? What exactly do we see here?

The photographs had been sent by Umlauff to the museum. The pictures confront us with the commercialized, objectified, and anonymous animal body in a stage of transformation that has rarely been visually documented. Surely the apes were commodities offered for sale. Bodies like these can be used for taxidermy as long as the skin is good.[55] The photographs offer a view into an historically obscure circumstance: A surplus of animal bodies, held in cold storage, waiting to be "naturalized" and relocated to a museum collection. Waiting for the separation of skin from body and flesh from bones; a transformation to a new number and a new status, a number that will move them from trade to science, from commodity to specimen.

Notes

1 The letters from Rowland Ward Ltd. may be written by another person, but they are always signed by Rowland Ward, and thus appear as a personal correspondence between Jägerskiöld and Ward.

2 For a thorough understanding of Walter Rothschild's importance as a zoologist and collector, see Miriam Rothschild's biography *Dear Lord Rothschild. Birds, Butterflies & History* (1983).

3 "Schritt für Schritt professionalisierte sich das Unternehmen Umlauff. Es gab zwar seine Herkunft aus der Unterhaltungsindustrie nicht auf, belieferte jedoch zunehmend bildungs-bürgerliche Institutionen, universitäre Forschungs-einrichtungen sowie natur- und völkerkundliche Museen. Diese doppelte Ausrichtung durchzieht die Umlauff'sche Geschichte" (Lange 2006: 12).

4 Letter from Rowland Ward Ltd. to L.A. Jägerskiöld 23.06.1906.

5 To elucidate the link between big game hunting and the collections in natural history museums, and to demonstrate how wild fauna became commercialized already in the field, a collection of hides and bones as well as the relics of a baby elephant Jägerskiöld bought from Swedish Magnus Leyer in Rhodesia in 1912, will be analysed in Chapter 4.

6 Berättelse rörande Göteborgs Musei Zoologiska Afdelning för 1909. *Göteborgs Museums Årstryck* 1910, 33.

7 Letter from L.A. Jägerskiöld to Rowland Ward Ltd. 26.06.1905.

8 Letter from Rowland Ward Ltd. to E. Lomberg 17.02.1904.

9 Letter from Rowland Ward Ltd. to L.A. Jägerskiöld 30.01.1906.

10 Letter from L.A. Jägerskiöld to Rowland Ward Ltd. 17.02.1906.

11 Letter from Rowland Ward Ltd. to L.A. Jägerskiöld 19.02.1906.

12 Letter from Rowland Ward Ltd. to L.A. Jägerskiöld 23.06.1906.

13 Letter from L.A. Jägerskiöld to Rowland Ward Ltd. 26.03.06.

14 Letter from Rowland Ward Ltd. to L.A. Jägerskiöld 29.06.1906.

15 "I was the first white man who has systematically hunted the beast, and who has at all penetrated to its haunts" (Du Chaillou 1861: 388).

16 Telegram to L.A. Jägerskiöld from Rowland Ward Ltd. 27.09.1906.

17 Telegram to Rowland Ward Ltd. from L.A. Jägerskiöld 28.09.1906.

18 Berättelse rörande Göteborgs Museums Zoologiska afdeling för år 1906. *Göteborgs Museums Årstryck* 1907, 14–18.

19 "Köpet måste vid jämförelse med hva, som ungefär samtidigt på annat håll betalats för motsvarande djur, betecknas som synnerligen billigt". Berättelse rörande Göteborgs Museums Zoologiska afdeling för år 1906. *Göteborgs Museums Årstryck* 1907, 18.

20 "Mitt första större tiggeri gällde en gorilla. Det var en praktpjäs, som en stor välkänd Londonfirma bjöd mig för £ 250. Jag telegraferade: "I am collecting money". Det behövdes i alt 5000 kr." (Jägerskiöld 1943: 347). Purchase of the gorilla was procured by donations from E. Bratt, D. Broström, Aug. Carlson, Carnegie & co, Rob. Dickson, H. Grebst, H. Hartwig, J.A. Hertz, Fr. Holm, C. Lyon, K. Kindal, C. Wijk, Hj. Wijk, A.O. Wilson, I. Wærn.

21 "... afdelningen är den enda, inom hvilken museiföremål verkligen framställas i större skala." Berättelse rörande Göteborgs Museums Zoologiska afdeling för år 1906. *Göteborgs Museums Årstryck 1907,* 26.

22 Letter to L.A. Jägerskiöld from Rowland Ward Ltd. 05.10.1906.

23 Hilmer Skoog's journal 1904–1924, archive number 368.

24 Photo GNM _226_1.

25 Letter from L.A. Jägerskiöld to Rowland Ward Ltd. 23.12.1905.

26 Letter from A. Girtanner to L.A. Jägerskiöld 07.05.1907. This modest sum must have paid for the taxidermy work alone.

27 Berättelse rörande Göteborgs Musei Zoologiska Afdelning för 1906. *Göteborgs Museums Årstryck 1907,* 18.

28 The General Catalogue reads: "Ma.ex. 618 Inköpt gen Girtanner från Amerika vildskjuten (?)".

29 Hilmer Skoog's journal 1904–1924, archive number 368.

30 "I don't think I ever would have to write such an unfriendly letter, but we can only use first-class". Letter from L.A. Jägerskiöld to A. Girtanner 30.01.1907.

31 Letter from L.A. Jägerskiöld to A. Girtanner 30.01.1907.

32 Letter from A. Girtanner to L.A. Jägerskiöld 24.02.1907 and 13.03.1907.

33 Letter from R. Bauzer to L.A. Jägerskiöld 03.02.1907.

34 Ma.ex. 571. Skeleton Coll.an. 4571.

35 "Den redan förut inköpta schimpanshanen från Gabun har under året af konservator Skoog färdigpreparerats och är ett vackert bevis på hans skicklighet samt står i utförande ingalunda efter den af den världsberömda firman Ward stoppade, 1906 inköpta gorillan." Berättelse rörande Göteborgs Musei Zoologiska Afdelning för år 1907. *Göteborgs Museums Årstryck 1908,* 19.

36 Hilmer Skoog's journal 1904–1924, archive number 368.

37 "Hos Hilmer Skoog", *Göteborgstidningen* 04.12.1916.

38 Letter from L. A. Jägerskiöld to Rowland Ward Ltd. 03.12.1906.

39 Letter from Rowland Ward Ltd. to L.A. Jägerskiöld 06.12.1906.

40 The camera was used on scientific and hunting expeditions from as early as 1860 to document and memorize nature: "Just as photographers drew on the skill of the taxidermist to overcome their cameras' technical shortcomings, taxidermists drew in turn on the photographer to provide them with an appropriate model of realism for their displays (Ryan 2000: 206– 207). Photographs were not only presentations of facts, but actors in societal and cultural negotiations, as Østgaard Lund and Berg claim with reference to the making of Norwegian national polar heroes (Østgaard Lund and Berg 2011: 9). Photography mediated the heroes' adventures to the public and directed people's attention to events following in the wake of the expeditions. Photographs were extremely useful in raising money for new adventures, documenting facts, and advertising (Larsen 2011: 13). As Tagg has demonstrated, at the end of the nineteenth century photography was also used to portray criminals, the mentally ill, orphans, and poor people, and functioned as a means of surveillance (Tagg 1993 [1988]).

41 The most famous manifestation of photography as an aid to good taxidermy was the Akeley camera.

Carl Akeley invented a motion picture camera that was patented in 1916 and used by the U.S. Army during the First World War (Bodry-Sanders 1996: 143–146).

42 "För att belysa djurens lefnadssätt i det fria, fåglarnas bobyggnad m. m. hafva en del starkt förstorade fotografier anskaffats. Äfven af sådana djur t. ex. noshörning, elefant m. fl., som museet ej på länge kan tänkas förvärfva hafva stora fotografier anskaffats." Berättelse rörande Göteborgs musei zoologiska afdelning för tiden 1:ste Juni –31:ste Dec. 1904.
Göteborgs Stadsfullmäktiges Handlingar 1905, 35, 8. Photographs and slides taken by Hilmer Skoog, David Sjölander, and Victor Hasselblad between 1913 and 1927 were exposed in the bird displays in the new museum building in Slottsskogen (Hedqvist 2009: 185).

43 Photographs GNM _229_1, GNM _228_1, GNM _226_1.

44 This means that M. Rothschild was incorrect in writing that the gorilla was "collected for him (Rothschild) in the Cameroons" (Rothschild 1983, illustration 63).

45 "Sie blieb in der Rolle des Händlers statt in der einer Institution. Dies begünstigte eine Systematisierung ihres Figurangebots und eine serielle Produktion für Ausstellungszwecke – auch ohne eigenes Museum" (Lange 2006: 84).

46 "Auch Tierbilder waren in diesem Kontext Bilder vom Menschen" (Lange 2006: 99).

47 "Es ist mir zwar bekannt, dass Sie in Ihrer grossen und reichhaltigen Sammlung bereits diverse Gorilla haben, trotzdem aber möchte ich nicht verfehlen, Ihnen auch dieses Exemplar anzubieten, das erstklassig ist". Letter from J. Umlauff to L.A. Jägerskiöld 14.06.1929.

48 Letter from Rowland Ward Ltd. to L.A. Jägerskiöld 30.06.1905.

49 GNM 6953_4.

50 GNM 5376_1.

51 "Seit der Gorilla Mpungu 1876 länger als ein Jahr lebend in Berliner Aquarium gehalten werden konnte, wurden die friedlichen und menschenähnlichen Züge der Tiere betont" (Lange 2006: 96).

52 http://www.br.de/radio/bayern2/sendungen/kalenderblatt/1311-gorilla-mpungu-europa-erster-100.html. Retrieved 14.12.2013.

53 http://en.wikisource.org/wiki/Popular_Science_Monthly/Volume_18/January_1881/Popular_Miscellany. Retrieved 14.12.2013.

54 "Gorillan använder trädgrenar till försvar, men dess fruktansvärdaste vapen är dock armen enbart. Ett enda slag av en gorillanäve kan fälla en motståndare stendöd till marken. Säkert är den retade gorillahanen en fruktansvärd uppenbarelse, där han kommer gående genom skogen upprätt på bakbenen, trummande med nävarna på sitt tunnliknande bröst och utstötande hemska vrålande ur det öppna gapet, vars hörntänder ej stå tillbaka för rovdjurens. Händer og tänder äro också gorillans vapen mot de fientliga skogsdjuren." "Tjocka släkten", *Göteborgs Aftonblad*, 02.02.1924.

55 Information from taxidermist Christel Johnsson 31.01.2013.

Macaca tonkeana (Meyer, 1899), female, 22 years

Provenance: Central Celebes

Captured: Pipikoro (landscape/county south of the Koro River)

Collected/ bought: Lemo in Koelawi

(landscape/county north of the Koro River)

Born: 1916

Dead: 19.12.1938

Put down by: Veterinary Ottander

Collected by: Walter Kaudern

Mounted by: David Sjölander 1943

Owner: Gothenburg Natural History Museum

Condition: Good

Entry in General Register: 1938-7467

Entry in Register of Foreign Mammals: 29.11.2004, Ma.ex. 1097

Entry in Collectio Anatomica: 7077

Location in the museum: Mounted hide on display in the Mammal

Hall; skeleton in the Bone Cellar

Captured : Monjet the Monkey   2

I n late December, 1938, a female Tonkean macaque called Monjet (the name means 'monkey' in Malayan), taken ill by measles a couple of days earlier, was put down by the veterinary, Ottander. The very fact that the monkey was euthanatized by a veterinary, rather than killed by a bullet, tells us that the animal had been living in surroundings quite different from those of her natural habitat. Unlike so many of the monkeys and apes taken to European menageries and zoos prior to the invention of a remedy for tuberculosis, this one did not die of this disease. Monjet had spent most of her life in the taxidermy room of the Gothenburg Natural History Museum. She was the first living creature, man or animal, to move permanently into the building, and this makes her a singular and significant case. When looking at Monjet in a broader context she appears as but one small creature in a vast multitude of exotica sent to Europe during the second half of the nineteenth century up to the Second World War. Yet, Monjet is unique because she lived almost her entire life in captivity, within the museum, where her mounted skin is now on display.

We know her year of birth, the date of her death, and fragments of her life as a museum monkey. Several stories emerged from her encounters with contemporary journalists, so she must have been well known by the city's residents during her lifetime. Although she died more than seventy years ago, Monjet is still remembered by people who met her in the museum, and she features in numerous narrated recollections. Moreover, there are surviving photographs showing the little monkey in her museum home: being caressed, drinking beer, and sitting together with another, much smaller live monkey. Of all the animals displayed in the Mammal Hall, Monjet is among those that have evoked the greatest public attention, and she is perhaps the star of them all. But unlike the bull walrus or the African bull elephant, which caught the visitors' attention as stuffed specimens, it was as a living creature that Monjet fascinated her audience. Her biography begins in Pipikoro in Celebes, in the middle of the First World War , and ends in a glass case in the Gothenburg Natural History Museum in the middle of the Second World War.

There is a striking contrast between Monjet's active and intense presence in the museum during the eighteen years she lived in the taxidermy room, and the stuffed Monjet. Her displayed mounted body is perhaps easily ignored as one moves through the exhibition. The little black monkey sits on the floor of glass case nr. 37 in a corner of the Mammal Hall – a position that makes her look considerably smaller than if she had been mounted upright. Maybe the taxidermist chose this position because Monjet is always shown in sitting position in the photographs depicting her. She lacks the visually striking attributes of other monkeys, such as the baboon's impressively grotesque buttocks or the colobus monkey's beautiful, lustrous fur. However, when one looks more closely at her face, one is puzzled by the vivid expression in her

*Fig. 20. Monjet sitting in the middle of the first row
in the glass case seen in the background.*

PHOTO: ANDERS LARSSON, GNM.

eyes. The glass eyes of mounted animals usually intensify the experience of looking at still life. Monjet's eyes enhance her personality and individuality.

When I first saw Monjet in the glass case, she looked to me as monkeys do mostly – as an average monkey – and an average monkey is what she is. My impression of Monjet's 'monkiness' was later confirmed when I read Dario Maestripieri's book *Machiavellian Intelligence. How Rhesus Macaques and Humans Have Conquered the World*. He writes:

If people were asked to think about a monkey, they would probably form an image in their minds that is either that of a young chimpanzee or a composite of all the monkeys and apes they've seen at the zoo or on TV. This imaginary 'average' monkey would probably look a lot like a rhesus macaque. (Maestripieri 2007: 7)

Or, perhaps, like a Tonkean macaque.

Macaques are the typical monkeys of the Old World, native to Asia and Northern Africa, and comprise about four different groups of species (Maestripieri 2007: 10). They are very competent animals, agile in trees and on the ground, good swimmers, and well capable of protecting their group. Macaques are featured in the entertainment industry, incorporated into zoos and amusement parks, and are frequently used in laboratories all over the world. This involvement in various globalised human endeavours makes the macaques the most widely distributed genus of non-human primates. The Tonkean macaque has a dark brown to black coat, lacks a tail, and has a pink rump. The female is slightly larger than the male, stands at 60 cm, and weighs around 10 kg. Voilà, Monjet!

The movement and transformation of both matter and meaning are essential to all the animal biographies in this book, but the contrast between a living creature and a mounted specimen is most evident in the case of Monjet. This chapter juxtaposes the two aspects of the monkey Monjet: as a pet and as a museum object. Monjet's biography unfolds as a history of a monkey worthy of remembrance, and contributes to the little known cultural history of the monkey in Sweden. A closer look at her case reveals that she was not wholly unique; her biography includes live primate species other than the Tonkean macaque. Importantly, because monkeys prefer to live in habitats and climates very different to those of Gothenburg, not only the monkeys themselves, but also their diets became an issue when discussing Monjet's biography. Finally, the chapter explores the emotional element in the collective story of natural history museums.

Monkey Business

On the famous Hereford Mappa Mundi, dating from around 1300, the Scandinavian peninsula is divided in two compartments, yet only Norway is named. While "(t)he figure of a hermit with his staff occupies the more northerly of the two compartments," the other contains a figure "of an ape squatting" (Bevan and Phillott 1873: 158, 159).[1] Why should Sweden or Suecia be symbolized by an ape, in Latin 'simmia'? Bevan and Phillot suggest that the monkey might represent the Turks who are placed close to Scandinavia on the map, and that the animal alludes either to the Turks' habit of eating apes and other despised "unclean" flesh, or maybe to the pygmies "who were supposed to tenant the extreme north, and whom Paulus Jovius describes in his letter to Clement VII as resembling apes" (Bevan and Phillott 1873: 159). I mention this because the Hereford map is a particularly early example of a visual representation of a monkey in a Nordic country.

The cultural history of the monkey in the Nordic countries seems to have been initiated with the word for the animal before the manifestation of the physical animal itself. The word 'api' is mentioned already as early as in the older Edda, where it refers to a fool and not to the animal; yet monkeys have often been portrayed as fools. Evidence of live monkeys in Sweden comes later than evidence of the word, and refers most probably to singular animals kept as pets at manors, or to animals travelling with troupes of jugglers in the summer. In Swedish texts, the monkey was referred to as 'apinia' in the feminine. The Swedish historian John Bernström, who has studied animals in medieval Sweden, explains the use of this form by the fact that female monkeys are easier to handle than males, and that the females therefore must have been the more frequently kept of the sexes.

Portrayals of monkeys before the Reformation indicate that the first monkeys brought to Sweden were tailless, and it seems reasonable to conjecture that they were Berber macaques. Long-tailed monkeys were first seen in Sweden after the Portuguese had reached India by sea, Bernström suggests. These monkeys were called 'markattor' (guenons) in Swedish. This word is never mentioned in medieval texts. 'Markatt' is a distortion of the Dutch 'meerkat', which derives from the Sanskrit word 'markaṭa' meaning monkey. 'Markattor' is the genus *Cercopithecus* and endemic to sub-Saharan Africa (Bernström 1956: 171–173). Pippi Longstocking's monkey Mister Nilsson, is a guenon.[2]

Historically, the monkey has triggered aversion in humans (Impelluso 2004: 198). This antipathy has been strong; indeed so fierce, that in Western culture the monkey has been associated with evil, the devil, and original sin (Bernström 1956: 172). In western religious art the devil has been portrayed as a monkey. However, as Impelluso points out, artists have also painted and sculpted monkeys as caricatures of humans engaged in activities such as playing cards, playing musical instruments, and even painting (Impelluso 2004: 198). A German snuff-box dated 1755–1760 from the Gilbert collection, on display at the Victoria and Albert Museum, offers a neat example of how artists have used monkeys to mock society. The external face of the lid is decorated with birds, painted in a naturalistic way. The inside of the lid, seen only by those invited to have a pinch of snuff, reveals a scene of animals dressed like people: "A monkey in a wig administers an enema to a prostate cat."[3]

Along with parrots, parakeets, and canaries, monkeys became much-coveted pets among the upper classes in the 18th century. Prominent naturalists such as the French Comte de Buffon (1707–1788) and the Swedish Carl von Linné (1707–1778) kept monkeys. Linné, who became skilled in raising monkeys, acquired his animals from the court.[4] Monkeys were the most popular exotic pets among the Parisian social elite, desired for their rarity and their association with the global trade, like foreign goods such as coffee (Robbins 2002: 141). During the 18th century, exotic animals were shipped to France via the same routes utilized for the slave trade. Louise E. Robbins observes that, even if exotic animals took on different meanings as commodities, scientific objects, pets, and symbols, the monkey moved constantly between these categories. Monjet's biography expresses this polysemy; she moved from being a wild animal to become a pet valued for her good temper as well as for her scientific value, and finally she was transformed to a species example and put on display.

Animals take their position in the human imagination in various ways and by different means: rhinoceroses, elephants, and polar bears impress with their magnitude; monkeys and apes through their ability to mimic us and to present our peculiarities in a distorted manner – they are at once both moving and repulsive in their anthropomorphic otherness. Their biological similarity to humans has made them vulnerable to cruel exploitation in the name of science. In Christian art, the monkey represented the devil; later we placed the monkey in the hell of experimental science.

The monkey as pet in Western culture is moreover a paradox, because this extremely agile animal has always been a chained pet. Robbins writes that 18th century Parisians intending to buy a pet monkey always looked for animals "that were *très doux* (very gentle) or *très-privé* (very tame) no doubt because they had a reputation for bad behaviour", or young animals that could still be tamed (Robbins 2002: 131). People don't like pet animals that bite, and the history of wild pet animals shows that one way to eliminate this problem is to pull out the teeth. John Sorensen observes that aggressive pet monkeys also have had their teeth removed, and that monkeys and apes, when kept in captivity, and especially when kept isolated, develop psychological problems (Sorensen 2009: 72).

Monkeys in Gothenburg

The glass case containing Monjet is one of four displayed monkeys, apes, and prosimians in the Mammal Hall. As I read the labels, I became increasingly surprised at how many of them had died in Gothenburg. The four monkeys on the shelf above Monjet had all died in this city between 1867 and 1905, as had the baboons that flank her. Three of the four Hapalines, each of a different species, also expired in Gothenburg; in fact the museum has acquired the majority of their displayed monkey and ape specimens from institutions, menageries, and local private donors.[5] This made me wonder how frequently monkeys have been in Sweden's most prominent shipping town in the period 1860–1940. Perhaps having a pet monkey was not as unusual in this period as it was when Astrid Lindgren wrote her famous books about Pippi Longstocking and her monkey, Mister Nilsson, in the early 1940s? Or was Astrid Lindgren inspired by stories of seaborne monkeys when she created the character of Mister Nilsson? Mister Nilsson is the link between Pippi and her father – the sea captain who became a "negro king" on an island somewhere in the Pacific, and the monkey is a link between the tidy and regular daily Swedish life, and the exotic, unexpected, and wonderful. In her lifetime, Monjet had also been a reminder of another world and another nature.

In the 18th century, monkeys were kept in orangeries, heated winter storage houses for citrus trees and other delicate plants. Ethnobiologist Ingvar Svanberg explains that Linné kept his beloved Diana monkey *(Cercopithecus diana)* in the warmest of the orangeries in the Botanical Garden during the harsh Uppsala winter. The little female black and white monkey slept on the beam that supported the ceiling, and the room's temperature, according to Linné, was as high as that of a sauna (Svanberg 2007: 66).

Around 1900, the citizens of Gothenburg could see monkeys in Cuneo's menagerie, one of the larger travelling collections of live exotic animals (Svanberg 2010: 118). Trädgårdsföreningen, the oldest botanical garden in the centre of Gothenburg, founded in 1842, has also housed monkeys and apes. As late as the end of the 1960s, a chimpanzee was living in a cage in the Tropical House, a department of the Palm House. It is said that the animal urinated on people.[6]

Various exotic animals including monkeys, alligators, and fish have also been kept in the city's Maritime Museum, Sjöfartsmuseet Akvariet. From 1933, the museum building was equipped with an "ape room" in the Tropical Department. Monkeys were displayed together with other tropical species, including snakes, alligators and lizards; animals acquired from sailors who had kept them on board as pets and souvenirs. Because of their foul smell, the monkeys were soon moved to a separate room. Species mentioned in the Maritime Museum archives are guenons, capuchin monkeys, bonnet macaques, and other unspecified macaques. Probably there had been monkeys in the maritime collections before 1933, since the new museum building was fitted with a separate room for them. Evidence shows that the visitors could see alligators from the 1920s until the beginning of 2000. A cast of the famous alligator, Smiley, that lived in the Maritime Museum for many years, is today displayed in the Gothenburg Natural History Museum. The monkeys in the Maritime Museum were put down in 1970 because of new animal welfare requirements.[7] The Natural History Museum has received monkeys and apes from each of these three institutions: Cuneo's menagerie sold their dead monkeys to the museum, while

Trädgårdsföreningen and the Maritime Museum gave theirs away for free.

During my search for live monkeys, other animals appeared as well. Then, as today, exotic animals were sequestrated by customs officials. After 1945 the quarantine office of Gothenburg was relocated to Ringön, but quarantine arrangements seem to have been based on a flexible approach to regulations. Quarantines could be established in museums, and even in private homes. In the 1950s and 1960s sequestrated animals were brought to the Gothenburg Natural History Museum. In June 1959 a tame five month old ocelot arrived in the free port of Gothenburg. The cat had been transported from Ecuador in a cage stored on the captain's bridge. Originally a gift from the "Banana King" Folke Anderson to his staff in Gothenburg, the ocelot was instead taken to the Natural History Museum to be kept in quarantine for four months.[8] It was not intended for the collection, and its destination after the quarantine period is unknown. A kinkajou cub, named Laban, arrived with the same ship as the ocelot, and was also lodged in the museum. Laban was given to Botaniska Trädgården in 1961, and apparently it lived in the taxidermy room for about two years. In the 1960s chief museum taxidermist Björn Wennerberg and his wife Ruth rescued wounded birds in a specially furnished unit in their home. Even exotic animals had a short stay in the taxidermy room before being transferred to the Wennerbergs.[9] These examples indicate that exotic animals circulated outside the realm of regular trade, but still within the limits of legitimate organizations such as the customs and the Natural History Museum.

Fig. 21. A male macaque sitting in its cage before being taken
to the museum and chloroformed. The monkey had been kept
in captivity for many years in Mölndal.

Some seamen kept their monkey shipmates after they had returned to Gothenburg.[10] In 1929 a photographer took a picture of a male macaque. It is a very sad scenario. The animal is caged and huddles on the branch of a tree trunk. A rope is fastened between the trunk and the branch. One side of the cage is made of wired fence. The monkey seems to be very frightened, and it had reason to worry. The photographer was taxidermist David Sjölander, and the monkey had been donated to the Gothenburg Natural History Museum, where it was chloroformed on 21 January, 1929, then flayed and skeletonized. The skull and the body were preserved in alcohol.[11] The photograph was probably taken just before the monkey was transported to the museum, somewhere in Mölndal, an industrial town south of Gothenburg. Here it had been kept in captivity for many years. How did this macaque end up in Mölndal? Probably it was one of the monkeys taken to the city by sailors.

Other evidence of monkeys in Gothenburg pops up in 1931, when a vivid dispute pro and con keeping wild birds in cages in Slottsskogen raged in the local newspapers. Would this be cruelty to the birds or not? A gentleman named Barthold Lundéus insisted strongly upon wild animals' capability to be tamed and live happy lives:

A wild animal can be tamed. And many of these animals, indeed maybe the majority, will be extremely faithful to their keepers, so they will not leave them for anything in the world.

These animals are no longer wild, but completely tame, he claimed. As an example Lundéus points at his "little editorial monkey" (redaktions-apa) Sissi that, he insists, *"is just as tame as professor Segerstedt's two dogs"*. Professor Segerstedt was one of his adversaries. Sissi, Lundéus writes, is just as loyal and faithful, and follows him and obeys him like a dog. *"But nevertheless she is born and captured wild in an Indian jungle. The gods know that she was as wild and shy as an animal can be when I got her eight years ago pretty directly from there."* [12] After two or three weeks the monkey had became so tame that she voluntarily came and sat on his knee, Lundéus claimed. Sissi, we learn, was another of the Old World's monkeys kept as a pet in Gothenburg, coming to town about a year after Monjet had arrived. One had been taken from Indonesia, the other from India.

In January 1936, a pinché monkey of the species *Leontocebus oedipus*, endemic to Colombia, was placed in the glass case reserved for the display of new objects. The present label explains that the monkey had died in Gothenburg the previous year, and had been given to the museum by one Gunnar Enemyr. The pinché monkeys belong to the *Callitrichidae* family, together with marmosets and tamarins. They have claws or claw-like structures on all fingers and toes except for the big toe, which is thumb-like and bears a nail (Sanderson 1957: 49). Small, with a body length of about thirteen centimetres, dexterous and with a shiny, rich coat, they were occasionally brought to Sweden by sailors.

In his press report, curator Paul Henrici seized the opportunity to teach the public how to feed these tiny creatures that so often, even

Fig. 22. Marmosets and pinchés. In the middle a Leontocebus oedipus, *dead 1935, to its left a* Hapale jacchus, *dead 1912 and in front a young* Callitrix sp., *dead 1924. The animals died in Gothenburg and were given to the museum by private donors. The ape to the right on top is a* Mystax rosalia, *dead 1893. Marmosets are usually gentle animals when kept in captivity, but bite if handled against their will, especially if a stranger is watching* (Walker 1964, I: 439).

PHOTO: ANDERS LARSSON, GNM.

when loved and cared for, languished and died. "They suffer from the climate, people say".[13] But it was the diet, not the climate, which killed them. Henrici's advice was to give them fruit, bread and all kinds of vegetables, but also insects: in summer, flies, larvae, grasshoppers and the like, and in winter, mealworms bought at the zoology store in Larmgatan. He also recommended that monkey owners give their animals 3–4 daily drops of 'Jectoral,' a dietary iron supplement available from the pharmacies. Henrici concluded: "And all of you who now sorrow for your ailing marmosets, will pretty soon be able to thank the good adviser in your hearts."[14] Pinchés cannot live solely on a vegetarian diet, since they need proteins found in insects, spiders, and fruit, but mostly in meat (Sanderson 1957: 66; Walker 1964, I: 439). A close look at their teeth reveals their dietary habits: "They have a set of teeth, and especially canines, that would, if enlarged, put any other mammal to shame, and they can wreak havoc on a full-grown crow" (Sanderson 1957: 66).

How many families in Sweden of the 1930s, with its high rate of unemployment, could afford to feed their pets a special diet, or even spend extra money on a pet in the first place? The majority of pet animals, such as dogs and cats, were fed leftovers. A scarcity of sufficient human nutrition did not favour special and expensive diets for monkeys, nor did the Nordic kitchen in which vegetables were few, especially during the long winter season, and most often cooked: onions, cabbage, potatoes, carrots, and other root vegetables.

Celebes

The Swedish ensajn,[15] Mr. O. Strandling, whose acquaintance I made during my stay in Soerabaja (now Surabaya) in February, has a very strong claim on my gratitude for his indefatigable efforts to do assignments for me in Java since we met. He made the magic lantern pictures, and for three months he took care of a monkey which I was especially keen on bringing home alive, let alone lots of other favours he did for me. (Kaudern 1921, I: 13 –14)[16]

The little monkey Walter Kaudern mentions here, in the preface of his travelogue "I Celebes obygder," was Monjet.[17] Zoologist Walter Kaudern, his wife Teres and their two sons Walter and Sven, four and three, left neutral Sweden on 5 December, 1916, with the steamer Baltic, bound for Australia, via Java.[18] Their destination was the interior of Celebes in Dutch East India, today Sulawesi in Indonesia, a region that until then had been visited by few (perhaps no) Europeans (Haraldson and Källgård 1997: 144). To depart from Sweden by sea at a time of the year when the weather is usually very rough, and while the world was at war, was a hazardous, even desperate move. The first entry in Kaudern's diaries from the Celebes expedition reads:

The end of 1914: Ever since I, had been schemed, to put it mildly, in a peculiar way from the position I had held in the years 1913–14 at the vertebrate department at Riksmuseet … I nourished a constant desire once more to be allowed to go on a new expedition, and to the full respire in freedom.[19]

In 1914, Kaudern was already an experienced traveller, having undertaken collections in Madagascar during 1906–1907 and 1911–12, and he wanted to continue his studies on the island. The Great War, however, forced him to change his plans, since the French authorities denied him entry to the island. Instead, he had to look for another territory, one administered by a neutral country, and with underexplored fauna. In 1916, when he at last gained permission to travel in Dutch East India, and with great difficulties had scraped together enough money for the expedition, he was unemployed. Moreover, he suffered from an arsenic intoxication because of his work in the museum – arsenic was used as pesticide in the zoological collections – and was on the verge of emigrating:

I began seriously to plan to use my last resources to simply get away as an ordinary emigrant. At the end of 1915 my brain was infested by God knows which plans, one wilder than the other. … I felt ever more clearly inside me that I, by all circumstances, had to leave this country that had turned me into a complete anti-patriot. I would soon hate all and everything. No one would in any way offer a helping hand.[20]

On 9 March, 1917, the family arrived at Dijko in North Celebes, and they continued the next day to the mining town of Goeroepahi, Kaudern on horseback, Teres and the boys in palanquins. In the central part of the island Kaudern would conduct a threefold program of fieldwork. His aim was to collect "a material as varied both anatomically and systematically as possible of the mammal species that are so characteristic of Celebes, these are *Babirussa* or the North Sulawesi barbirusa, the *Anoa* or the miniature buffalo and *Cynophitecus* or Celebes' close to tailless monkeys" (Kaudern 1921, I: 8).[21] The monkeys were a focus of the expedition, and the collections of Gothenburg Natural History Museum contain skins from sixteen examples of *Macaca nigra*, the Sulawesi black macaque, and five examples of *Tarsius spectrum*, the spectral tarsier. Monjet is the only example of Macaca tonkeana in the Gothenburg Natural History Museum, and it is worth mentioning that as a specimen, she was first registered in the museum's General Register as *Cynopithecus tonsus*.

Parallel to zoological, and botanical, mapping and collecting, Kaudern conducted ethnographical studies and collected artefacts from indigenous people. The expedition became a turning point in Kaudern's academic career (and initiated Monjet's biography!), because his knowledge of Celebes' native people would come to be of greater importance to his future career than his zoological fieldwork on the island.[22] Although Celebes' ethnography became Kaudern's main focus in the expedition, when the family travelled back to Sweden via Suez late in 1920, a small macaque had joined their group.

As for Monjet's life in Celebes, the General Register says that Kaudern acquired the monkey in May or June of 1918, but in a letter to Jägerskiöld, Kaudern writes that he obtained it in early July.[23] At approximately one and a half years old, Monjet was still an infant when she became Kaudern's property. In July 1918, the Kaudern family had their camp in the village of Lemo in the Koelawi district.[24] Koelawi is north of the river Koro, Pipikoro; Monjet's homeland lies south of

Koro. The monkey had been captured by local people when she was between six and nine months old, then tamed and trained by them. Kaudern writes that people in Koleawi, the district where he conducted his main fieldwork, were eager to obtain goods from the outside world. Clothes made of bass,[25] domestic utensils, eggs, rice and all kinds of animals, live and dead, were bartered for white, red, and blue cloth, thread and yarn, glass pearls and boxes with a mirror inserted in the lid, combs, toothpicks (!), salt, lime and *gámbir*, "a dainty" made of the leaves from *Uncaria gambir*, and chewed with lime and pinang or betel nut (Kaudern 1921, I: 119–220, 227).

Kaudern bought Monjet from a male resident in the village of Kantewoe in Pipikoro, so the opportunity to acquire foreign goods must also have attracted people from the neighbouring district to seek the Kauderns. I wonder what was the trade equivalent of a tame, infant (Monjet still had her milk-teeth), female macaque? Was she worth a box with a magic mirror, or was she traded for glass pearls, a stereotypical transaction between Europeans and Old World natives in which items of value are obtained for next to nothing?

The monkey must have been of considerable value, since Kaudern arranged for it to be looked after for three months in Java, while he was busy elsewhere (Kaudern 1921, I: 13). But what kind of value did the animal represent? Neither his diary, nor "I Celebes obygder" provide detailed information about how and why he obtained a live little monkey. In his diary on 3 July, 1918, Kaudern wrote in the margin: "Bought this day a half grown monkey from a Kantewoe man. The monkey should have been captured a bit south of the Koro River."[26] He

Fig. 23. In September 1918, Walter Kaudern with his family crossed Koro via this impressive bridge. Monjet had joined the family two months earlier, and travelled with them. The bridge was 47 meters long and made of rotting wood. It hung about 12 meters over the water. It could carry a weight of approximately 1,000 kg. "It's hard to deny that to the beginner it feels a bit hazardous when he, like a rope-dancer, comes out on the narrow bridge that for each step swings, high up in the air and with the thundering Koro under" (Kaudern 1921, I: 346).[27] To cross the bridge must have been an easy task for Monjet.

PHOTO: VÄRLDSKULTURMUSEET.

113

notes that he has written a letter to his mother and that the boys made drawings with crayons. But he also writes that his wife Teres did not feel well that day, and two days later she had a violent attack of malaria, her temperature rising to 40.9 degrees centigrades.[28] She continued to suffer from malaria after the family had returned to Sweden.[29]

To travel with a live animal presupposes very different dispositions than travelling with dead specimens, practically as well as mentally. And Monjet was not the only live animal acquired by the Kauderns. Kaudern mentions that before the expedition set out for poorly accessible Koelawi, they could sometimes care for about ten live animals, all with different diets. One of these animals was the tiny and extremely shy spectral tarsier, a nocturnal animal weighing just 80–100 grams and rarely surviving contact with humans. Tarsiers feed on tree snails, tree frogs, and insects (Sanderson 1957: 48).[30] A python could live on a chicken every second week, Kaudern writes, and the tortoises lived quite well without food. The tarsier, on the other hand, demanded a variety of insects (Kaudern 1921, I: 85–86).

A tame parrot was also adopted by the Kauderns. It was Teres Kaudern's special pet, and she intended to bring it back to Europe. The bird suffered an unhappy fate, however. During the voyage back to Sweden, Monjet, who was extremely jealous of the parrot, managed to snatch the bird and, quick as lightning, tear off its head.[31] The expedition, with its mix of human beings and animals, was an unusual sight in the inlands of Celebes. When a military patrol met them on 16 June, 1919, in an area with only one European inhabitant, the soldiers were more than surprised to see a white man travelling with his wife, two small sons, servants, a black monkey, a parrot, chickens, and a row of porters (Kaudern 1921, II: 218–219).

Was Monjet brought to Sweden alive not simply because she survived in captivity, but also because she, after two years in the company of the family, had obtained a kind of pet status? Kaudern travelled with his sons, and even if Monjet had been collected mainly for scientific purposes, the tame monkey would most probably have become their playmate. The travelogue gives no answer. Is there no resonance of Monjet in Kaudern's travelogue from Celebes simply because details of family life and their developing feelings for a monkey were too private to be inscribed in this rather dry and matter-of-fact journal? The closest we get to Monjet in Celebes, when reading Kaudern's travelogue, is a photograph showing "our tame, black monkey from Central Celebes who was brought to Sweden alive" (Kaudern 1921, I: 226).[32] Referring to the monkey as "our" points at a conception of her not necessarily as a pet, but surely as a creature who lives within the family circle.

Sven Kaudern, Walter Kaudern's youngest son, has confirmed, however, that Monjet obtained the status of playmate and pet. He was 97 years old when I spoke with him in the spring of 2011, and even after ninety years he retained vivid memories of their simian friend. When they were in Celebes, Monjet had a little kennel attached on the top of a pole, where she slept during the nights. One morning she was unable to come down from her kennel because her fingers were stuck in it: she would always pluck on things and tear them to pieces. She was also very skilled at catching things. The family could throw a raw egg at her, and she would catch it without breaking the shell. She was also able to pluck off an egg's membrane with her tiny fingers. Monkeys watch us and imitate our behaviour. Monjet observed and imitated how the boys washed their faces, by taking water in her hands to clean her face, and afterwards pouring the water from the washbowl. And the boys learned from Monjet, too. Sven told me that she taught them to climb trees. Monjet could laugh, and she also had a very special way of moving her ears, one ear independently of the other. By watching her carefully, Sven learned to move his ears in the same way, an art he told me that he still performs. The father and mother studied the monkey, the monkey studied the boys, and both the boys and the monkey studied each other – a loop of learning and pleasure to all of them.[33]

The monkey itself is mostly absent from Kaudern's diaries as well as his travelogue, but we do learn about Monjet's homeland, Pipikoro. She came from a wild mountainous region, where peaks range up to 3000 metres. Rivers and waterfalls traverse the land, interspersed by deep, verdant valleys. Pipikoro is located approximately 1500 metres south of the river Koro, which is one of two water systems in Koelawi. Before the Dutch colonised Celebes, Pipikoro was an independent principality, well protected by natural boundaries. From this rainforest highland, Monjet travelled first to the coast, where she suffered in the warm, humid climate.[34] She left Baoe Baoe on the island Boeton, bound for Java, in June 1920, and from Java she was taken to Suez. The next destination was Hamburg, and from here the family continued the journey northwards to their final destination, the arid urban landscape of the windy and chilly Gothenburg. Monjet arrived on 15 February, 1921, after a journey that involved passages on seven different ships.

A Very Special Offer

For Kaudern, three years in the rainforest was a strain, both mentally and physically, but this was offset by his ability to conduct his studies without the interference of superiors. His longing for freedom fuelled his efforts to finance the Celebes expedition, and hence his departure from Celebes was also a farewell to the freedom he had enjoyed there. In June 1920, when his family waited for a ship from the island of Boeton, Kaudern wrote the final lines in his travelogue: "We looked forward to the voyage home with little enthusiasm. In Celebes we had been living in many years so to speak outside the world, unaffected by what happened there. Our work had been our only, totally absorbing interest, we had for so long been wandering our own ways in the wilderness, that we stepped back when we had to return to civilized land." [35] Coming home would involve, among other things, securing the material collected, and creating a safe lodging for Monjet. In January 1921, Leonard Jägerskiöld received a letter from Walter Kaudern, written from the Suez Channel, which included a rather special question: could Jägerskiöld house a live monkey in his new museum?

B. Brother![36]

As we are now approaching home I will mention that I bring a live monkey from the area of Kantewoe in the innermost region of Central Celebes and I would like to see that in the future it might live as long as it will suit it. According to Byttighaver (?) in Rotterdam no monkeys from Central Celebes are identified for sure.[37] I have had the beast in captivity since the first days of July 1918, and with lots of inconveniencies and I have been able to transport it no less than five times on Dutch ships only thanks to the Dutch authorities' great courtesy.

Regrettably we have no zoological garden and I will never place it in the terrible monkey house at Skansen. I cannot keep it myself, but I have been thinking of the possibility that you perhaps could house one or two animals in your new museum. The beast is by the way very tame.

If you can and will receive the animal in case it gets home alive, I will be grateful if you would be kind enough to write me some lines and direct them immediately to the address S/S Lygnern* Hamburg, because we are supposed to arrive there one of the first days in February: Our ship will not sail further, so I have to find another shipping opportunity from there and home. If you will take the monkey, I will do as much as I can to carry it with me, otherwise I will leave it in Hamburg.

So far the voyage home has proceeded well, but from now on remains probably nothing but cold and drizzly weather. We have already had a slight foretaste of the nasty cold.

Best greetings to you and the museum from your friend Walter Kaudern
*The Swedish Consulate[38]

In January 1921, Jägerskiöld was deeply engaged in finishing and furnishing the new museum building. Was a live monkey the most suitable gift for his new museum in this hurly-burly? Did Kaudern's offer arouse enthusiasm? Jägerskiöld answered on 3 February, 1921:

Brother W. Kaudern

c/o S/S Lygnern. Schwedisches Consulat

I was happy when I received your letter from Suez. Your collections arrived already in Sept. and we have unpacked them and done what I found suitable to prevent any damage.

What pertains to the beast, I am willing to receive it and to try to keep it alive. I don't have any special arrangement for live animals and it must be kept in a cage in full view in the taxidermist's studio. I suppose that he can live of carrots, bread, some fruit every now and then, and also some milk. This on the condition that the animal is complete and pretty and that it, when it passes away, will be given to us.

Welcome! I suppose that you will pass by Gothenburg because you must definitely order your collections in person and give the facts about them.

In haste with warm regards

Your friend J. A. Jägerskiöld

P.S I don't know if there is quarantine for monkeys, but I suppose it will be fixed.[39]

It is amazing that Jägerskiöld should agree to house the monkey on the condition that it was "complete," as if he had forgotten that he was negotiating for a live animal and not a dead specimen. He was also expecting a male animal; as demonstrated in Chapter 1, good specimens were male specimens. Kaudern, on the other hand, is clearly negotiating acceptable terms on behalf of his family pet, to secure her life, albeit one of cramped captivity.

From my conversation with Sven Kaudern I learned that it was of great importance to his father to transport Monjet to Sweden alive, and to procure good housing for her, since the family had no place to live when they returned to Sweden, and were hence unable to keep her. Kaudern was happy that Jägerskiöld accepted a monkey in the new museum, and he did not have any reservations against Jägerskiöld's supposition for the monkey's diet. Carrots, bread, and milk – vitamins, carbohydrates and minerals – reflected the basic elements of a then new, optimal diet for school children in Oslo, introduced in the 1920s, and later known as "Oslofrokosten" or the "Oslo Breakfast" (Lyngø 2003). The main constituents of this meal were in fact fresh milk, bread, and some kind of fresh fruit or more often – and cheaper – a carrot, very different from a simian diet rich in fruit, berries, and the fresh greenery of central Celebes. Macaques feed on a wide variety of plants and animals. Monjet, who lived in captivity for eighteen years, must have been fed on protein rich food.

Kaudern's letter of 12 February, 1921, helps us to better understand why he worked so hard to bring the monkey to Sweden:

S/S Lygnern Hamburg 12 February, 1921

B. Brother!

Thank you very much for your letter, and I'm happy that the monkey is allowed to end up in Sweden. Everywhere people have tried to buy it, also on Celebes, Java, Cebu, and not to mention here in Hamburg, where someone even tried to steal it. The zoological garden here is very eager to get it (i.e. Hagenbeck Tierpark), but since I, with lots of bother and inconvenience succeeded to bring it here, and not for business purposes, I will of course take it all the way home. Of course the intention is to give it to you for free, the only condition is that it is allowed to live.

At last I have succeeded in arranging that we can all travel by ship to Gothenburg. If all goes well, we will arrive around noon on Monday, because we leave with the S/S Tatti this evening by the Kiel Channel and the beast is allowed to come with us.

Otherwise I carry with me very little of zoological interest except for two skulls with small antlers of wild or rather feral buffalo from Luzon. The antlers are, however, curved in a way that differs from those of a common domestic buffalo.

Well done that you have arranged my collections to be looked after. I think that much of them may never be packed again because skins and similar material should remain in Gothenburg. Stockholm Högskola has few possibilities to keep such kinds of collections.[40] … I think, anyway, that the geological collections are so big that neither Stockholm Högskola nor Gothenburg should complain.

You can therefore take the necessary steps to take the monkey ashore. My best greetings to E. Nordenskiöld, S (?) and you yourself from the friend Walter Kaudern.[41]

Jägerskiöld did not reply, so we do not know exactly what steps were taken to care for Monjet, but it is reasonable to assume that she, like the animals mentioned earlier in this chapter, was kept in a kind of quarantine in the taxidermist's workshop. Kaudern's letters make clear that she had value as a rare specimen, and that her rarity would guarantee her a lifelong stay in the museum. It is also obvious that Kaudern and his family cared for Monjet, although Kaudern never refers to her as their pet. On the contrary, he refers to her as the 'odjur,' which means 'brute' or 'beast'. Kaudern's writings demonstrate, however, that he very rarely, if ever, expressed his positive feelings in elated words. Malaria, fever, otitis, toothache, destroyed feet, insect bites, the Spanish influenza (the epidemic raged in Celebes in the autumn of 1918), pouring rain and steep, slippery paths, lack of horses and porters, birthday and Christmas celebrations – all were registered in a scientist's sober and precise style. When I said to Sven Kaudern that there was very little about Monjet in his father's diaries, he replied dryly: "And about us!"

Two episodes from the journey from Hamburg to Gothenburg confirm that Teres Kaudern and Monjet were close – and that a pet monkey is far less easy to control and handle than a dog. In Hamburg harbour the family was transported in a smaller boat to the ship that would take them to Gothenburg. From this vessel they would enter the ship by a ladder. Mrs. Kaudern was supposed to be the first to climb onboard, carrying Monjet in her arm. But Monjet, being the better climber, boarded first, to the attending mariner's great astonishment. Later during the voyage, the family was invited to the captain's saloon. The gentlemen had cognac, while Teres Kaudern sipped liqueur with Monjet on her lap. The second time the captain filled the glasses, Monjet snatched her glass and emptied it.[42]

A Home in the Museum

During the year 1921, the exhibition halls, workshops, offices and store rooms in the new building in Slottsskogen were gradually filled with skins, bones, glass jars, and tubes, stuffed animals, chemicals, protocols, and books. The transfer of the zoological collections was a well planned process, and is described by Leonard Jägerskiöld in his annual report for the year 1921:

Some animals or minor collections have occasionally been transported to the new museum in Slottsskogen, and during the autumn the entire taxidermy workshop has been moved there. At the same time the taxidermist began to work in the new studio, which has proved to be very suited for its purpose. In this way the final transfer has been initiated. It will be carried out with our own employees. To this end a small van has been bought, and one of the custodians has been trained as a chauffeur. Judging from the experiences so far, the removal promises to continue without difficulties and the result should be a joy to behold.[43]

I suppose Monjet was lodged in the museum immediately after the Kauderns had disembarked in Gothenburg, on February 1921. In the following two years, Walter Kaudern was busy systematizing his zoological collection from Celebes. He was working in the new museum building, which means that he and Monjet spent their days in the same place, but not necessarily in the same room.[44] Monjet was, however, not confined to the taxidermy room, nor did she lack company. The Kaudern boys came regularly to see her, and Rolf, a friend of theirs,

Fig. 24. Transport of collections to the new museum building 21 April, 1922.

GNM_920_1.

sometimes took her to his home, and for walks in Slottsskogen, where he let her climb the trees. [45]

The little monkey had moved into the taxidermy room even before taxidermist Hilmer Skoog, who began to use his new workshop in the autumn of 1921. Skoog was the first of the staff to start working in the museum; Monjet kept him company from the first day. Chained to the

CHAPTER 2

Fig. 25. Walter Kaudern systemises his collections from Celebes.
Barbirusa crania are lined up on the bench. It took him close to two years to
complete this work. 17 October, 1921.

PHOTO: HANDELSTIDNINGEN. GNM_941_2.

wall outside her little wooden box, she could observe dead birds and
animals being dissected and flayed, and skeletons being taken apart,
cleaned, and reassembled. Bodies were modeled, and skins shaved and
arranged on mannequins. Strange smells from chemicals and dead
bodies activated her nose.

Fig. 26. Hilmer Skoog in the taxidermy room 1922.
Monjet is sitting to the right outside her kennel.
There are two barbirusa manikins and a barbirusa skeleton.

PHOTO: K.A. UTGÅRD, GÖTEBORGS DAGBLAD. GNM_921_1.

Did she ever react with fear, aggression or perhaps joy towards a mounted animal? Monjet was present when Skoog mounted two of the *Barbyrousa celebensis* that Kaudern had collected in Celebes: did the process evoke any excitement in her? She also witnessed Skoog when he made the preparatory works for the walrus mount in winter and spring 1927.

CHAPTER 2

Fig. 27. Friends. Hilmer Skoog and Monjet, 8 May, 1924.

PHOTO: COUNT ARVID POSSE. GNM_952_2.

Skoog and Monjet had a close relationship: how did she react when her friend Skoog never returned from his summer holiday that year? In her new living quarters, there were interesting things to learn and imitate. After she had been observing a man plucking a bunch of pigeons he had shot in Slottsskogen, Monjet freed herself from her chain during the night. The next morning a wounded falcon that had been cared for in the taxidermy room, was found with all its feathers picked off.[46]

Monjet apparently thrived in her museum home, but what did the taxidermists think about their cohabitation with a Tonkean macaque? A close up photograph showing Hilmer Skoog and Monjet leaves little doubt that Skoog had warm feelings for the monkey. There is no evidence of David Sjölander taking the same strong liking to the animal. A year after Skoog had died and Sjölander had become Monjet's 'master,' the monkey had been able to free herself from her chain, while she was alone in the studio. The result was disastrous.

Under the heading "Monjet's work of destruction 22/8 1928" Sjölander has listed the acts of vandalism that ruined the collections from his summer fieldwork in Lapland.[47] During the night of 21 August, Monjet bit, tore and crushed whatever she found, using her deft hands to open and empty various boxes and bottles. It is impressive to read how ravaging a small monkey can be. She had crushed 33 eggs from seven different bird species. She had spoiled 10 kg of potato flour, painting, turpentine, different oils, soda, and a dish brush. Even worse, she had ruined film and photographs from Lapland, destroyed a camera, and torn an umbrella and a bag to pieces. Pens, pencils and Indian rubber were in bits and pieces. She had smashed a bottle of red ink. There were broken bones and crushed craniums; the microphone of the telephone was bitten into pieces, and she had poured tea, sugar and spikes on the floor, and eaten half a kilo (!) of dried plums. Letters and notes written by Sjölander were also destroyed. Obviously the doors to the studio had either been left open, or Monjet had been able to

Fig. 28. Monjet as taxidermy.

PHOTO: ANDERS LARSSON, GNM.

open them herself. She had inspected the formalin box in the hall and visited the bathroom, opened a window and thrown out glasses, bitten the rubber tube off the compressor. She had also entered ichthyologist Orvar Nyblin's office and ruined some of his glass tubes, and spread cocoa across the floor. "Satan's animal! – Satans djur!" was Sjölander's conclusion. But Monjet was not thrown out of the studio.

The episode explains why the museum always called Teres Kaudern to come and calm down Monjet, when she succeeded in escaping her chain. It also helps us to understand why the staff, during those times when Monjet's rich fur was at its best, would suggest that maybe it was time to put her down.[48] When this happened in 1938, Sjölander was in charge of the taxidermy, and in 1943 he completed the very expressive mount of her skin.

Friendliness

Monjet appealed strongly to children and young adults. In his travelogue, Kaudern reports an episode which indicates that a tame monkey was also a source of fascination to people in Celebes. On the evening of the second of June, 1919, the expedition had reached an outlying village called Kelei, near the coast. The Kauderns had hardly unpacked their luggage, before the 'pasanggrahan,' or guesthouse, was surrounded by curious children and women. The two fair boys especially held their attention, but the most interesting creature was Monjet. Young people in the village soon became aware that the monkey, at the sight of people, would open her red throat widely, smile, and display all her impressive teeth. The creature was far less dangerous than she appeared, however. During the family's stay in the village, at least twenty people competed daily to pay her attention, offering her titbits and playthings. "The monkey was completely dizzy of joy for the kindness that seemed never to come to an end".[49]

The scene in Kelei represents the essence of Monjet's behaviour towards children: friendliness. Her wide grin, displayed when befriending people, became famous. A rhesus macaque shows its teeth to tell another macaque that it is its subordinate. The human smile often functions similarly (Maestripieri 2007: 144). Baring teeth, a signal of submission in many other macaques, is used by Tonkean macaques as an affiliation signal, often during play. A study of Tonkean macaques has produced data which indicate that their pattern of behaviour is different from that of rhesus macaques. One striking difference is what they communicate by silent bared-teeth display, that is, what we would call a smile: "It does not express submission, but signals peaceful intentions on the part of the emitter, and serves to initiate affinitive interactions" (Thierry, Anderson, Demaria, Desportes and Petit 1994: 106).

Monjet's behaviour can also be recognised in what is said about Tonkean macaques being especially manipulative and destructive, for instance, escaping from their enclosure or using objects in play. They are also quick to learn (Thierry, Anderson, Demaria, Desportes and Petit 1994: 107). All evidence indicates that Monjet showed her teeth to people as a non-aggressive act. "His only company in his workshop is a monkey who shows its teeth to express pleasure with its visitors, so they say," reported the newspaper Ny Tid in October 1922.[50] *Göteborgs*

Handels- och Sjöfartstidning's journalist, who had met Skoog and Monjet six months earlier, had likewise interpreted Monjet's "smile" as an expression of joy. He describes Monjet as a rarity: the only other example known in Europe was a stuffed specimen in Berlin, although it was not confirmed that the two belonged to the same species. The journalist was more impressed by Monjet's good character than by her rarity, however. The monkey "has a noble character, and a mild and sensitive temper, and she immediately invites us to like her. She is caressed a lot – and with a peculiar, smiling laughter she shows that she is happy to receive so many proofs of kindness. She is truly an extraordinarily nice monkey, and she has learnt a genuine human sense of humour, playfulness – and thirst."[51]

The children were part of Monjet's flock: on one occasion, Rolf, a friend of the Kaudern brothers, had taken Monjet to his home. The children and the monkey were playing in a sandbox, when a neighbour, who disliked the monkey, approached with his dog in order to bully her. As the dog approached Monjet, she grabbed its back and lifted it up. "Now you see how dangerous Monjet is!" Rolf exclaimed.[52]

Monjet also befriended other animals. On the occasion of the Gothenburg Natural History Museum's centenary in 1933, *Göteborgs Morgenposts Söndagsbilaga* published an extensive article on museum taxidermy and research.[53] At this time, there were two live monkeys in the taxidermy room, Monjet and a guenon. The details of the latter's life as a museum monkey are difficult to elucidate. The guenon is only documented in three photographs, two taken in the summer of 1931, and the third in October 1933. The monkey is not mentioned in any of

Fig. 29. Monjet and the guenon.

the newspaper articles in the museum scrapbook for 1931, but Monjet is referred to in an article dated the first of March, 1931 (as a baboon).[54] If there had been two monkeys present in the taxidermy room that day, the journalist would certainly have mentioned it. In June 1935, the little guenon was still in the taxidermy room along with Monjet. The journalist's task was primarily to report on wounded birds of prey that had been taken to the museum for treatment. On this day there were as many as five, but the monkeys were obviously better material for a story. A photograph shows taxidermist Ekström leaning down while Monjet is grooming his thick hair. The guenon is sitting next to Monjet, watching Ekström.

Once more Monjet is described as a kind animal that likes to groom the taxidermist's hair, while her little companion seemed to be the opposite, as he both bites and shows his tongue to people.[55] It certainly did not live long in the museum, and it is only vaguely remembered by some members of staff. The monkey has not been registered in the General Register. Most probably this means that it must have been moved to another place while alive. My suggestion is that it could have been donated to the Maritime Museum, since this museum received a male guenon in 1937. His name was Jocke, and is described as "green and a little sly," maybe of the species *Chlorocebus sabeus*, called Green monkey or 'vitgrön markatta' in Swedish.[56] Among the six monkeys mentioned by name that were placed in the Maritime Museum that year, Jocke was the only one with a cunning temper.

Today the museum holds Monjet's mounted skin and her skeleton. It is unclear what happened to the body immediately after her death; the General Register says that skin and torso are preserved in spirits, and that the skin was later mounted and displayed. Most probably the body was skeletonized at the same time – and a glance at the bones shows that this was done rather roughly. The teeth are remarkably healthy, especially when Monjet's age and her consumption of sweets and beer, are taken into consideration.

The Director Orvar Nybelin did not find it worth mentioning in the museum's 1938 annual report that their mascot of eighteen years, a female macaque, which, owing to her rarity, had been so coveted in Hamburg that people had tried to steal her, now had been incorporated in the collections as the museum's first example of *Cynopithecus tonsus*

Fig. 30. Monjet's bones.

PHOTO: ANDERS LARSSON, GNM.

(Matschie), today *Macaca tonkeana*. The only item specifically mentioned among the nineteen animals added to the collection of foreign mammals that year was the skin and skull of an African male lion, obviously a welcome piece that when mounted should replace "the by now aged example, which to the public hardly justifies its title as king of the animals."[57] Even an annual series of *Dicrostonyx hudsonius* (Hudson Bay collared lemming), summer and winter skins of *Mustela erminea* (ermine) and a skin of *Phoca hispida* (ringed seal) are reported by Nybelin, and must have been considered of greater scientific value than the remains of Monjet.

A Paradoxical Animal

Monjet was brought from Pipikoro to Gothenburg as a result of a frustrated scientist's expedition to Celebes. She was a collector's item, and in the correspondence between Kaudern and Jägerskiöld she is referred to as the beast. After her death her remains became part of the collection. However, conversations with Sven Kaudern, the newspaper articles and the "obituary" written by Paul Henrici and published on the occasion of Monjet's debut as a mounted animal in 1943, as well as the photos of Hilmer Skoog and Monjet, lead me to believe that she clearly had acquired the status of a museum mascot, and to some people had been a dear pet. Henrici calls her "our good and faithful friend, who lived a quiet and apparently thriving existence."[58] During her eighteen years in the museum, Monjet had shaken hands with members of the Bernadotte house, the Swedish royal family, and in their honour, and for her own pleasure, she had emptied several bottles of pilsner, procured by 'landshövding' von Südow, the highest state representative in the county. Monjet also participated in the celebration of Jägerskiöld's seventieth birthday on 12 November, 1937, dressed in a uniform splendidly decorated with orders from the firm Buttericks.[59]

On 4 April, 1943, the newspaper *Göteborgs Handels- och Sjöfartstidning* reported: "The Monkey Monjet on a place of honour in a cabinet for novelties."[60] While she lived, about 15,000 schoolchildren visited the museum annually, and she was known by all people in Gothenburg and half of Western Sweden, *Göteborgs Handels- och Sjöfartstidning* concluded.[61]

Even if Monjet had met with prominent people and with hundreds of schoolchildren, she cannot be considered a celebrity animal of the kind exemplified by Clara, the Indian rhinoceros, Jumbo, the African elephant, and Knut, the polar bear. There is a striking difference in the ways in which these historical celebrity animals were presented to the public, compared to Monjet's backstage existence in the taxidermy room. Clara, Jumbo, and Knut were showcased in order to attract an audience. In the years 1741–1758, Clara toured France, Bohemia, Denmark, Austria, Switzerland, Poland, Italy, and England in order to make her owner, the Dutch sea captain Douwemout Van der Meer, a wealthy man. She was so famous that she became a powerful figure and motif in applied arts as well as in fashion (Ridley 2004).[62] Jumbo was literally the biggest animal celebrity in the London Zoo from 1865 until 1882, when he was sold to P.T. Barnum to feature in his circus Barnum and Bailey. This transaction provoked heated protests, and has been interpreted as a "national trauma" in which the sale of the popular Jumbo became a symbol of Britain and the Empire handing over influence to its rival, the United States (Ritvo 1987: 232). The polar bear Knut, who died in the Berlin Zoo in 2011, was an animal celebrity typical of our time. Knut was seen as, among other things, the incarnation of collective concerns about climate change, and its impact upon the polar bears' dwindling habitat (Flinterud 2013).

It is also tempting to compare Monjet with another famous animal, the Saint Bernard Barry, preserved and displayed in the Bern Natural History Museum. This dog lived in the Saint Bernard Monastery between 1800 and 1812, and died in 1814. Barry was an extraordinarily clever rescue dog and became famous during his own lifetime. After his death, he was stuffed and displayed, because he had served man in an exemplary way. Today Barry is still famous, and is considered to be one of the most important mounted animals in the museum. The specimen is presented to the public as a dog of myth and matter. The mythical Barry draws people to see the mounted Barry (Nussbaumer 2000, Thorsen 2013).

Monjet's fame and agency shrank after her death, from being a monkey whose reputation had reached far beyond the museum, to be reduced to an anecdote transmitted by the staff inside the museum. To diminish an animal from a unique individual to a mere functional exemplar of a species erases its celebrity status. Contrary to the Saint Bernard in Bern, who features in the museum entrance hall, Monjet in the glass case has been utterly divorced from her life as an individual monkey. The bargain between Kaudern and Jägerskiöld was based on the Kaudern family's affection for Monjet, and Jägerskiöld's enthusiasm for collection. Unlike Clara, Jumbo, and Knut, Monjet' fate was, from the very beginning, to become a mounted specimen.[63] Before this could happen, however, she had to be kept somewhere. What could be more convenient than to lodge her in the taxidermy workshop itself?

Monjet has been naturalized in a double sense: first mounted, and then displayed without any indication of her domestication – the belt, the chain and the pilsner bottle – not even with a label explaining why she died in Gothenburg and not in her natural habitat. She used to meet children with one arm stretched towards them, gibbering as she showed her teeth. She moved her head from one side to the other while she "waved" with her ears. And when the children laughed, Monjet jumped with joy like a rubber ball.[64] Today you can hear schoolchildren exclaim "Creepy!" when they observe the glass cases containing the bodies of monkeys and apes.

Contrary meanings intersect within Monjet's little body. Her biography demonstrates that categories such as wild and tame, pet and specimen, individual and exemplar, subject and object, are permeable and interchangeable. We learn that natural history museums have harboured live animals, and so that the functions of museums and zoos, to some extent, have overlapped. Monjet's story is also one of many, illustrating that people obsessed with collecting dead animals may also care very much for live ones. It also reveals that small apes were less exotic in Gothenburg at the end of the 19th and the beginning of the 20th century than they are today. And finally, it reminds us that co-habitation with live monkeys is always a somewhat risky undertaking.

Notes

1 Thanks to Jørgen Sneis who made me aware of the monkey on the Hereford Mappa Mundi.

2 In the books Mister Nilsson is a guenon, but in the Pippi Långstrump movies the monkey is a common squirrel monkey, *Saimiris sciureus*, borrowed from a family in Stockholm. It did not behave properly, and was hence elimenated from the script and sent back to its family. http://sv.wikipedia.org/wiki/Pippi_L%C3%A5ngstrump. Retrieved 02.01.2014. The common squirrel monkey is native to the Amazon basin; the guenon is one of the Old World's monkeys.

3 The label reads: "Loan. Gilbert. 494–2008". The motive is an example of 'singerie' which means monkey trick, and in the *Encyclopedia Britannica* 'singerie' is defined as a "type of humorous picture of monkeys fashionably attired and aping human behaviour, painted by a number of French artists in the early 18th century".

4 For Linné's monkey pets see Svanberg 2007: 61–69.

5 Hapalines are marmosets, tamarines, and pinchés.

6 E-mail from Ellen-Marie Berggren Samuelsson, Trädgårdsföreningen, 24.05.2011.

7 E-mail and letter from Anna-Lena Nilsson, keeper of the archives, Sjöfartsmuseet Akvariet, Göteborg, 27.05.2011.

8 In 1948 Folke Anderson (1903–1968) founded several plantations in Ecuador and cultivated bananas and other fruits, and later he founded the Folke Anderson Fruit Trading Corporation. This was a success, and he was nicknamed "Banankungen", "the Banana King", in Swedish newspapers (Rydberg 2008: 119–120).

9 Interview with taxidermist Monica Silfverstolpe 31.10.2011; interview with taxidermist Erling Haack 31.10.2011; "Naturhistoriska fick pigg ozelot", *Göteborgsposten*, 13.06.1959; "Unikt djur på SAL-båt", *Göteborgs Handels-och Sjöfarts-Tidning*, 03.09.1959; "Den bölande tjuren", *Göteborgstidningen*, 17.04.1960; "Bananbjörnen Laban flyttar till Trägår'n", *Ny Tid*, 14.07.1961.

10 I have not found concrete evidence from families who have lived with seaborne monkeys in their homes. What I say about monkeys taken to Gothenburg by seamen and monkeys kept in private homes is based on information from the Maritime Museum, newspapers, protocols and labels in the Gothenburg Natural History Museum.

11 Ma.ex. 969, Coll.an. 5573.

12 "Ett vildt djur kan tämjas. Och många af sådana djur, ja, kanske de flesta, bli då sina vårdare så ytterst tillgifna, att de inte för allt i världen längre vilja öfverge dem. ...
Förresten vil jag i detta sammanhang påstå, att Sissi, min lilla redaktionsapa, *är fullt ut lika tam som någon af professor Segerstedts båda hundar*. Och minst mig lika tillgifven och trogen. Hon följer och lyder mig alldeles som en hund. *Men ändå är hon född och fången i en af Indiens urskogar*. Och guderne må då veta, att hon var så vild och skydd som ett djur gärna kan vara, när jag för en åtta år se'n först fick henne så godt som direkt därifrån." "Djursskyddsintresse under falsk skylt. En Röfvarehistoria i "Slaskröret"", *Vidi* 14, 02.04.1931.

13 "De tåla ej klimatet sägs det". "Nytt på Naturhistoriska museet", *Handelstidningen*, 12.01.1936.

14 "Och Ni alla, som nu sörja edra tynande silkesapor, skola rätt snart finna anledning föreligga att i edra hjärtan tacka goda råds givare." "Nytt på Naturhistoriska museet", *Handelstidningen*, 12.01.1936. Ivan Sanderson made an identical comment on the general ignorance of pinché monkeys and the

assumption that they are vegetarians: "People buy the poor little creatures, take them home, buy special cages for them, lavish care and affection on them, and then suffer profound frustration and distress when they die for no apparent reason" (Sanderson 1957: 66).

15 Ensajn is a degree in the Swedish Salvation Army assigned after seven years of successful service. http://svenskuppslagsbok.se/tag/ensajn/. Retrieved 22.07.2013.

16 "I alldeles särskild grad har svensken ensajn O. Strandling anspråk på min tacksamhet, ty sedan jag i februari under mitt besök i Soerabaja gjort hans bekantskap, var han outtröttlig i att uträtta uppdrag för mig på Java. Till mina föredrag på Java 1920 gjorde han skioptikonbilderna, och en apa, som jag var särskilt angelägen om att levande föra med hem, vårdade han tre månader för mig, mycket annat att förtiga" (Kaudern 1921, I: 12–13).

17 The title *I Celebes obygder* means *In unsettled Celebes*. The travelogue is based on Kaudern's diaries, here referred to as Dr. Kaudern's Celebes-expedition I, II, III. The diaries are in Världkulturmuseet.

18 Walter Kaudern (1881–1942) was a zoologist and geologist. After the expedition to Celebes, ethnography became his main field. In 1928 he was employed at the Gothenburg Natural History Museum's geology and mineral department, where he made geology displays.
In 1932, after Erland Nordenskiöld's death, Kaudern became responsible for the ethnographic department, and was appointed department head in 1934. Parts of the zoological material from Kaudern's Celebes expedition is in the Gothenburg Natural History Museum, parts in Stockholm's Högskola, today the University of Stockholm; the botanical material is in the National Natural History Museum, Naturhistoriska Riksmuseet, in Stockholm. The rich ethnographical material incorporates 3,000 objects bought by the Gothenburg Museum, and is today housed in Världskulturmuseet, the Museum of World Culture in Gothenburg. Kaudern published the ethnographical collection in the series *Ethnographical Studies in Celebes*, 6 volumes (Skottsberg 1943, Wassén 1942). Kaudern was also a skilled amateur painter. His oil paintings from Celebes are in Världskulturmuseet.

19 "Slutet av 1914: Allt sedan jag på ett minst sagt egendomligt sätt blivit bortintrigerad från den befattning jag innehade under åren 1913–14 vid Riksmuseets vertebratavdeling … närde jag en ständig attrå att åter få ge mig ut på en ny expedition och i fulla drag få andas in friheten". Dr. Kauderns Celebes-expedition I: 1.

20 "Jag började på allvar tänka på att med utnyttjande av mina sista ressurser helt enkelt ge mig av som vanlig emigrant. För övrigt genomkorsades min hjärna mot slutet av 1915 av Gud vet vilka planer, den ena vildsintare än den andra. … Under dylika omständigheter ingick år 1916 och jag kände allt tydligare inom mig att jag under alla omständigheter måste bort från detta land som numer gjort mig till fullkomlig antipatriot. Jag hatade snart allt och alla. Ingen ville på minsta sätt räcka mig en hjälpsom hand." Dr. Kauderns Celebes-expedition I: 4–5.

21 "… ett så rikhaltigt både anatomiskt och systematiskt material som möjligt av de däggdjursformer, som äro karakteristiska för Celebes, nämligen *Babirussa* eller hjortsvinet, *Anoa* eller den lilla vildbuffeln och *Cynopithecus* eller Celebes' nästan svanslösa apor" (Kaudern 1921, I: 8). The Gothenburg Natural History Museum has 62 examples of the

barbirusa collected by Kaudern, 8 of these are females; and skeleton fragments of 79 miniature buffalos collected by Kaudern.

22 An example of his anthropological studies is Walter Kaudern: Anthropological Notes from Celebes. *Ethnological Studies 4*, 1937.

23 Letter from W. Kaudern to L.A. Jägerskiöld 16.01.1921.

24 Kaudern and his family left Lemo in Koelawi 4 September, 1918, and initiated a three day journey to Kantewoe in Pipikoro (Kaudern 1921, I: 337–351).

25 After the Dutch had conquered the Central Sulawesi in 1905, the inhabitants' way of life changed completely. They had to convert to Christianity, and they gave up their traditional costumes, made of bass, for Western clothes (Amnehäll 1997).

26 "Kjöpta denna dag en halvvuxen apa av en kantewoe man. Apan skulle vara fångad något söder om Korofloden." Dr. Kauderns Celebes-expedition III: 48.

27 "För nybörjaren förefaller det onekligen litet äventyrligt, när han likt en lindansare kommer ut på den smala, för varje steg gungande bron högt upp i luften med den dånande Koro innunder" (Kaudern 1921, I: 346).

28 Dr. Kauderns Celebes-expedition III: 49.

29 Interview with Sven Kaudern (b. 03.08.1913) 31.05.2011.

30 A tarsier from Kaudern's expedition is displayed in a glass case together with the pinchés in the Gothenburg Natural History Museum.

31 Interview with Sven Kaudern.

32 "Vår tama svarta apa från Central-Celebes, som levande medförts til Sverige" (Kaudern 1921, I: 226).

33 Interview with Sven Kaudern.

34 "… hon plågades av det varma kustklimatet på Celebes, allt enl. dr. W Kaudern". (General Register 14851).

35 "Vi sågo fram mot hemresan utan någon särskild glädje. På Celebes hade vi under flera år levat så att säga utanför världen, oberörda av det, som där passerade. Vårt arbete hade varit vårt enda, allt uppslukande intresse, vi hade så länge vandrat våra egna vägar i vildmarken, att vi ryggade tillbaka, då vi måste åter till civiliserade bygder" (Kaudern 1921, II: 404).

36 "Beste Broder," "B. Broder"or only "B.B." meaning 'Best Brother' was the common way in which Swedish colleagues and friends addressed each other when corresponding in the 1920s, 1930s, and 1940s.

37 After Kaudern had decided to travel to East-India in autumn 1916, he made a preparatory journey to the Netherlands in July 1916, and visited some central institutions, among them the Natural History Museum in Leiden, the Amsterdam zoo and its aquarium and museum, as well as the Rijksmuseum. I suppose that Byttighaver was one of the persons he met during his stay in the Netherlands. Dr. Kauderns Celebes-expedition I: 10–12.

38 "B. Broder! Som vi nu närma oss hemmet vill jag nämna att jag medför en levande apa från trakten av Kantewoe i det indre av Central Celebes och att jag gärna skulle se at den for framtiden finge leva så länge den finner det passande. Från Central Celebes äro nämligen enligt Byttighaver (?) i Rotterdam inga apor med säkerhet kända. Jag har haft odjuret i fångenskap sedan de första dagarna av juli 1918 och med mycket besvär och endast tack vare de holländska myndigheternas store tillmötesgåande har jeg fått transportera den ej mindre än fem gånger på de holländska båtarne.

Tyvärr har vi ingen zoologisk trädgård och på Skansen i dess grässliga aphus kommer jag aldrig att sätta den. Själv kan jag ej bevara den, men jag tänkte möjligen att du i ditt nya museum kanske kunde inhysa ett eller annat djur. Odjuret är för övrigt mycket tamt. Kan och vill du taga emot djuret i händelse det kommer levande hem, så vore jag tacksam om du ville skriva några rader med omgående till mig under adrss s/s Lygnern* Hamburg, ty vi torde vara där någon av de första dagarna av februari: Vår båt lär ej gå längre, utan jag måste söka annan lägenhet därifrån och hem. Vill du ha apan, skall jag göra vad jag kan för att få den med mig, annars lämnar jag den i Hamburg. Hittills har hemresan gått utmärkt, men nu återstår väl intet annat än kyla och rysk. Vi ha redan fått en liten försmak av den hemska kylan. De bästa hälsningar till Dig och museet från vännen Walter Kaudern. * Svenska konsulatet." Letter from W. Kaudern to L.A. Jägerskiöld 16.01.1921.

39 "Brother W. Kaudern
c/o S/S Lyngern. Schwedisches Consulat
Jag blev glad när jag i dag fick ditt brev av från Suez. Dina samlingar kommo redan i sept. och ha vi packat upp dem samt vidtagit de mått o steg som jag ansåg lämpliga för att intet skulla förfaras. Angående odjuret är jag villig mottaga detsamma och försöka holda det vid liv. Någon särskild anstalt för levande djur har jag ej utan det måste då hållas i en synlig bur i konservators ateliern. Jag antar att han leva av morötter, bröd, något frukt då och då samt litet mjölk. Detta under vilkor att djuret är helt och vackert samt att det vid eventuelt dödsfall skänkes till oss.
Välkommen hem! Jag antar att du (?) tar vägen över Göteborg ty du måste nog absolut själv gå igenom dina samlingar hämma och ange facta om

dem.
I hast men med de bästa hälsningar vännen L.A. Jägerskiöld
P.S. Jag vet ej om det är karanten för apor men det ordnar sig väl antar jag. D.S."
Letter from L. A. Jägerskiöld to W. Kaudern 03.02.1921

40 Stockholm Högskola, today University of Stockholm, had granted Walter Kaudern 5,300 Swedish crowns to study the zoology of Celebes for a year. Dr. Kauderns Celebes-expedition I: 13.

41 "B. Broder! Hjärtligt tack för ditt brev, och det glädjer mig att apan kan få hamna i Sverige. Överallt har man velat kjöpa den, både på Celebes, Java, Cebu och ej minst här i Hamburg där man t.o.m. försökt sig på att stjäla den. Zool. Trädgården här är mycket angelägen om att få övertaga den, men nu som jag med mycket trassel och besvär lyckats få den hit och ej gjort det i affärssynpunkt vill jag naturligtvis föra den enda hem. Naturligtvis är det meningen att du får den gratis, endast med det villkor att den får leva. Sent omsider har det nu lyckats mig få det så ordnat att vi alla kunne fara med båt till Göteborg. Om allt går väl äro vi där vid middagstiden om måndag. I kväll resa vi nämligen med s/s Tatti via Kielkanalen och odjuret får följa med. For övrigt har jag ingenting vidare med mig i zoologisk väg med undantag av ett par skallar med granna horn av vild eller snarare förvildad buffel från Luzon. Hornen har emmelertid en hel annan svängning än vanlig tam buffel. Det var ju bra att du ställd om att mina samlingar blivit omsedda. Jag tänker att en hel del aldrig behöver packas in igen ty skinn och dylikt skola väl stanna i Göteborg. Stockholm Högskola kan ju knappast ha någon anledning att få med av sådana samlingar. … Jag tror dock att de zoologiska samlingarna äro så pass

stora att bade Högskolan och Göteborg ej skola kunna beklaga sig. Du får således taga de mått ock steg som en fordras för att taga apan i land. De bästa hälsningar till E. Nordenskjöld, S (?) ock dig själv från vännen Walter Kaudern". Letter from W. Kaudern to L.A. Jägerskiöld 12.02.1921.

42 Interview with Sven Kaudern.

43 "Enstaka djur eller smärre samlingar ha vid olika tillfällen flyttats till det nya museet i Slottsskogen och under hösten flyttades hela konservatorsverkstaden dit. Samtidigt förlades konservatorns arbeten till den nya atelieren, som visat sig synnerligen lämplig för sitt ändamål. Härmed är början till den slutliga överflyttningen gjord. Den kommer att företagas med eget folk. För ändamålet har inköpts en mindre lastbil och en av vakterna har upplärts till chaufför. Att döma av den erfarenhet som gjorts, lovar flyttningen att gå utan svårigheter och lyckligt." Berättelse rörande Göteborgs Musei Zoologiska avdelning för år 1921. *Göteborgs Museums årstryck* 1922, 44.

44 Berättelse rörande Göteborgs Musei Zoologiska avdelning för år 1921. *Göteborgs Museums årstryck* 1922, 46; Berättelse rörande Göteborgs Musei Zoologiska avdelning för år 1922. *Göteborgs Museums årstryck* 1923, 38.

45 Interview with Sven Kaudern.

46 Interview with Sven Kaudern.

47 "Monjets förstörelsesverk 22/8 1928".

48 Interview with Sven Kaudern.

49 "Apan var alldeles yr över förtjusning över denna velvilja, som aldrig tycktes ta slut" (Kaudern 1921, II: 184–185).

50 "Innan vi lämna museet måste vi göra en påhälsning hos konservator Skoog, som har ett ansvarsfullt arbete. Genom hans händer skola djuren vandra innan de komma ut till allmänt beskådande i museet. Bristfälligheter skola överses och där det fattas något blir det att göra en naturtrogen kopia, något konservator S. är förfaren i. Hans enda sällskap i arbetsrummet är en apa, som visar tänderna av förtjusning över att få besök, påstods det." D.Y.: "Bland vilda djur i Slottsskogen", *Ny Tid*, 21.10.1922.

51 "En så fin apa är i och för sig värd uppmärksamhet, men som vistas i hr (sic) Skoogs hägn har dessutom en fin karaktär och et vekt og känsligt lynne, som gör att man genast måste hålla av henne. Det kelas ochså rätt mycket med henne – och i ett sällsamt, grinnande leende visar hon sin glädje över alla ömhetsbetygelser. Det är en ovanligt rar apa, och hon har tillägnat sig en sann mänsklig humor, lekfullhet och – törst." "I zoologiska museets konservatorsatelier." *Göteborgs Handels- och Sjöfartstidning*, 24.03.1922.

52 Interview with Sven Kaudern.

53 "Naturhistoriska Museet 100 år", *Göteborgs Morgenposts Söndagsbilaga*, 28.10.1933.

54 "Fem mtr hög giraff konserveras. Konservator Sjölander berättar om sitt arbete", *Ny Tid*, 01.03.1931.

55 "Sjukhus för rovfåglar", *Morgontidningen*, 19.06.1935.

56 E-mail from Anna-Lena Nilsson, keeper of the archives, Göteborgs Sjöfartsmuseum, 29.11.2011.

57 "... för det nuvarande ålderstigna exemplar, som inför allmänheten knappast gör djurens konung rättvisa." Naturhistoriska Museet. Berättelse för år 1938. *Särtryck ur Göteborgs Musei årstryck* 1939, 11.

58 "... vår goda och trogne vän fört en lugn och till synes trivsam tillvaro". "Apan Monjet på hedersplats i nyhetsmonter", *Göteborgs Handels- och Sjöfartstidning*, 02.04.1943.

59 "Apan Monjet på hedersplats i nyhetsmonter",
    Göteborgs Handels- och Sjöfartstidning, 02.04.1943.

60 "Apan Monjet på hedersplats i nyhetsmonter",
    Göteborgs Handels- och Sjöfartstidning 02.04.1943.

61 "Apan Monjet på hedersplats i nyhetsmonter",
    Göteborgs Handels- och Sjöfartstidning, 02.04.1943.

62 The effect of seeing and studying a live rhino was
    imprinted on fashionable objects like porcelain
    and clocks and on hair fashion. Up to Clara's
    entrance in the world of celebrity animals,
    representations of rhinos copied the woodcut
    Albrecht Dürer made in 1515 of an Indian
    rhinoceros (Ridley 2004).

63 Jumbo was preserved. Knut has been mounted and
    will be displayed in Museum für Naturkunde in
    Berlin.

64 "Apan Monjet på hedersplats i nyhetsmonter",
    Göteborgs Handels- och Sjöfartstidning 02.04.1943.

Odobenus rosmarus rosmarus (L.), m; Atlantic walrus

Provenance: East Greenland

Collected: Rörö

Dead: 9 January 1927

Shot by: Anders Olsson

Collected by: Leonard A. Jägerskiöld

Taxidermy by: Hilmer Skoog and David Sjölander 1927–1928

Owner: Gothenburg Natural History Museum

Condition: Fairly good

Entry in the General Register: Ma.su. 651 1927- 4567

Entry in Collectio anatomica: 5026 (5026–5034)

Hide, skeleton, throat, heart, stomach, spleen, caecum, aorta, urogenitalia, suprarenal glands

Location in the museum: Taxidermy hide on display in the Whale Hall; skeleton in the Bone Cellar. Soft tissue parts and baculum stored with material from other mammals in a container.

Red-listed: The walrus on Svalbard has been completely protected from 1952.

Stranded : The Walrus from Rörön   3

In the autumn of 1926, a bull walrus left the drifting ice near the east coast of Greenland and began what was to be his last journey. Walruses typically follow the edges of coastal ice, and travel south when the ice begins to overtake their summer foraging grounds, but for some reason, this one headed southeast, crossed the Norwegian Sea and passed Iceland to his right. According to the newspaper *The Scotsman*, a walrus was seen in Scottish waters around the first of October, where it remained for a couple of weeks. In the middle of October, a walrus was observed from Sumburgh Head lighthouse on the main island of Shetland. One of the men on duty could clearly see the animal through his telescope, and particularly its large tusks. The next observation of a walrus was made by some fishermen on the island of Hisken, Norway, on 26 November. The following day, a walrus was seen from the island of Bømlo, east of Hisken. In early January the following year, H.R. Redeke, director of Zoölogisch Station der Nederlandische Dierkundige Vereeniging (the Biological Station at Helder), announced that a walrus had been observed and photographed from Holland on 11 November, 1927, at Fort de Harsens at Nieuwediep, close to the southern inlet of Zuider. Several attempts to catch the foreign animal during its visit to Dutch waters failed.

From Holland the walrus then turned north; but instead of swimming north–west (back towards its typical route), he headed towards the west coast of the Danish peninsula Jutland. At the end of November he swam into Hanstholm Bay, where he was harassed by local fishermen in motorboats, but escaped. At 10 p.m. on 5 January, 1927, the walrus came ashore between Grenen and Skagen, at the northernmost point of Jutland, where he was shot and wounded in the head. After this assault the walrus left Denmark, crossed Skagerrak to the east, and reached the archipelago north of Gothenburg, Sweden. On Sunday 9 January, a man discovered the huge animal lying on a skerry off the island of Rörö. The skerry happened to be about a hundred meters from the home of the region's most skilled seal hunter, Utter-Anders, who killed the walrus cleanly and efficiently with a single bullet to the head.

Walruses are agile swimmers and can reach a maximum speed of 30–35 km per hour. For longer distances the speed will be between six and seven km per hour.[1] It is likely that the walrus in question had followed a cold ocean current that runs from eastern Greenland southeast to Iceland and the Faeroes, then crossed through the warm Gulf Stream, and followed the cold southern current to Holland. From Holland this current turns northwards towards Jutland (Jensen 1927b). Subsequent observations of walruses indicate that other individuals have followed a similar route; some have been observed in southern Norwegian waters prior to turning north.[2]

Walruses have occasionally visited the coasts of Scotland and Norway, but this one distinguished himself to be exceptional. The arrival of a walrus so far south, in Holland, and the presumption that the

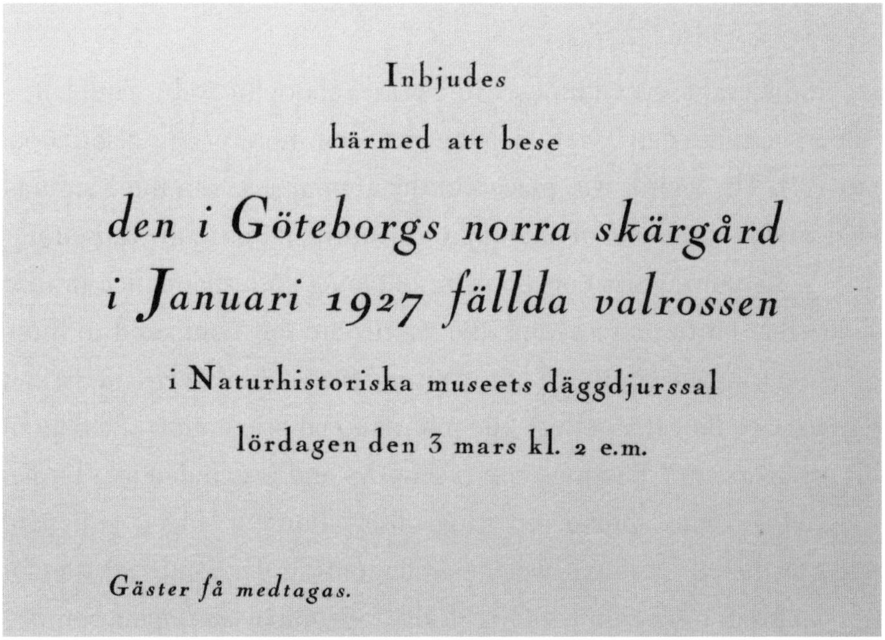

Inbjudes

härmed att bese

den i Göteborgs norra skärgård
i Januari 1927 fällda valrossen

i Naturhistoriska museets däggdjurssal

lördagen den 3 mars kl. 2 e.m.

Gäster få medtagas.

Fig. 31. Invitation to the see the walrus on the specimen's first day on display, Saturday 3 March, 1927, 2 p.m.

PHOTO: ANDERS LARSSON, GNM.

animal had visited four countries before it was killed in a fifth country, generated interest among scientists and naturalists, as well as among commercial agents in the field of natural history. This interest peaked in Gothenburg, where members of Biologiska Föreningen, the Biological Society, and other prominent citizens were solemnly invited to see the walrus mount for the first time on Saturday 3 March, 1928, at 2 p.m. The next day more than 1200 people visited the museum, and the number of visitors culminated in March.[3] The walrus's remarkable final destination by the seal hunter's cottage inspired the local newspapers to report the event several times, while stories about the walrus and Utter-Anders continued to be retold in the local community (Gustafson 2009: 159–161). The walrus was placed in the central space in the Mammal Hall, and became one of the public's favourite items in the museum.

The museum's copybook for outgoing correspondence in 1927 shows that entries concerning the walrus are few compared to those documenting the continual presence and flow of birds, eggs, nests, and leg rings – a flow that reflects the international orientation of a natural history museum.[4] The museum received – and responded to – numerous letters from abroad reporting observations of ringed gulls and other feathered creatures, welcome elements in the continual attempt to map avian migration patterns. Today, Scandinavia's *lingua franca* is English. In 1927, however, the museum corresponded with foreigners in English, German, Dutch, French, Italian, and Spanish. The diversity of languages reflects the wide geographical range of migrating birds, and the ways in which observations of ringed birds united scientists and bird lovers internationally.

The advent of the walrus aroused naturalists' attention for a relatively short time, but the walrus also inspired letters, articles, and photographs that shed light on the perceptions of a large and strange animal by scientists, naturalists, and the local public. The Gothenburg Natural History Museum archives contain twenty letters concerning observations of a walrus's movements beyond its natural habitat. The letters document various attempts to identify the species, and to deduce whether one or several walruses had been observed. The letters also describe the numerous ways humans have behaved, when confronted with such a strange, large and probably to many, also monstrously ugly animal.

"… but we would fain bespeak for it till the end goodwill from all men"

Whence in the Arctic range of its race came this strange visitor we cannot tell, and the further course of its long journey may remain a mystery; but we would fain bespeak for it till the end goodwill from all men. (Russel 1927: 99)

With this sentence, solicitor, amateur naturalist, and town clerk at Lerwick in the Shetland Islands, George W. Russel, ended his *Country Life* article on the walrus watching he and his friends had enjoyed from Hillswick on the west coast of the Shetland mainland. Russel's report on walrus watching expresses scientific curiosity paired with genuine sympathy for the beast, and offered an approach to outdoor life different from that of the many articles on foxhunting, dogs, and racing that *Country Life*, launched in 1897, featured during its initial years (Strong 1996).

The first observation of a walrus visiting the coast of Shetland in 1926 was made on the first of October. Next a walrus (perhaps a different one), was seen several times in the waters around the Shetlands during the next two months. It was a huge, scarred animal with tusks up to a length of fifteen inches or more, and most probably a bull walrus. Russel wrote:

The writer, with friends, had made a number of journeys by sea and by land, mainly along the Bressay shore, in an endeavour to fall in with an object of so much interest. (Russel 1927: 97)

There were few reported landings, however, hopes of localizing its movements and getting close to it vanished as the weeks passed. Then, on 28 November, Russel in the company of two friends sailed out in a small boat to finally make the animal's "acquaintance" (Russel 1927: 97). They observed it at close range for four hours. His description of the walrus's first appearance alludes to the world of shapeless amoebas rather than the splendour of arctic animals:

The appearance of the walrus at the surface, undisturbed was greatly different from our expectations. It lay so low in the water that, although it showed a considerable extent of its back, as well as the top of its head, the mass was featureless. Looked at from above, it rather had the appearance of a dirty bed-sack almost submerged, and a broadside view gave us an impression which, perhaps, was best described of a gigantic caterpillar out of its element. (Russel 1927: 97)

Russel then describes in detail how the walrus was feeding along a small skerry for hours, coming up to the surface to breathe about every two and a half or three minutes, keeping its head so low in the water, that the tusks were rarely visible: "Indeed, the uniformity of its movements that day when undisturbed was remarkable, and was extremely aggravating to the photographer" (Russel 1927: 98). Clearly the walrus's docile behaviour while foraging on mussels did not correspond

*Fig. 32. Walrus swimming in Scottish waters, resembling
"a gigantic caterpillar out of its element". Shetland 28 November, 1926.*

GNM_1588.

to the photographer's efforts to get a picture of the animal in an alert, erect posture. When the walrus was eventually disturbed, and "reared up in the water exposing its tusks and huge neck, throwing itself rapidly round towards the exit from its position, and making off", the poor photographer did not react in time (Russel 1927: 98). He was obviously out of luck, since the only photo taken of the walrus "with tusks well up and a considerable part of its neck showing, at a distance of three or four yards, [was] a failure, though the same exposure had been as quick as the light would allow" (Russel 1927: 98).

At the end of the day the company decided to test how closely they could approach the walrus by boat, and concluded that it was "so little watchful that we formed the opinion that we could have run the boat on top of it, had we wished" (Russel 1927: 99). Then men and walrus parted ways: "… the last we saw of it as we travelled homewards that evening was its head showing as it crossed the reflection of a dim glow in the western sky, ending a day we had keenly enjoyed" (Russel 1927: 99).

Tortured

A walrus in foreign waters was a marvel to humans – the walrus's last days were, however, far from marvellous. People assaulted the unfortunate animal in every locality it was observed. The first shot fired against a walrus during the autumn of 1926 reportedly occurred at Hillswick on the west mainland of Shetland, delivered by "a watchful Customs man who discharged a shot-gun in the immigrant's face from a few yards' distance" (Russel 1927: 97). When a walrus was swimming in the waters outside Schlicken, at Niewediep-Helder between 11 and 14 November, 1926, people first tried to catch the animal alive, then fired shots at it. The walrus subsequently left Dutch waters and headed north (Redeke 1927: 89). At the same time, by a peculiar coincidence, the Russian steamer *Garibaldi*, loaded with live animals, happened to sink at the mouth of the river Elbe after a collision. Before it became clear that the ship did not have any walruses on board, it was thought that the animal might have escaped from the steamer (Redeke 1927: 89).

The most gruesome attacks, however, took place in locations along the west coast of Jutland. Hunters without experience or knowledge of how to pursue such a large animal attempted to kill it using a pistol, a shotgun, a rifle, and even an iron stick.[5] This foolish cruelty evoked harsh feelings in Jägerskiöld, being both a devoted zoologist and an experienced gentleman hunter. Correspondence between Jägerskiöld and Danish zoologist and professor Adolf Severin Jensen (1866–1953) at the Danish Museum of Natural History in Copenhagen indicates that the cruelty inflicted upon the walrus had been too offensive to be ignored. Five days after the walrus had been shot at Rörö, Jägerskiöld, who learned about the physical state of the animal during the process of skinning and dissecting, wrote to Jensen:

Dear colleague.

Thank you for the information about *Torpedo marmorata*.[6]

I suppose that Danish newspapers have reported on the walrus that visited Skagen on the 7th or 8th this month. They filled him with shotgun pellets, wads, etc. in his eyes and elsewhere. Thank the Devil that the tortured and blind animal escaped. The marvellous thing was however, that he went off to Bohuslän and lay down on a skerry about a hundred meters from the cottage of the best seal hunter at Rörö. This was on 9 January. This man who is now fishing for herring, was at home, because it happened to be a Sunday, and with his Mauser he gave the walrus a quick and painless death. An old male with powerful tusks now belongs to the Gothenburg Museum. You may thus trust what the newspapers report on the walrus' presence at Skagen, which concerns your fauna. I would be happy if you could flog or at least brand the villains who mistreated the walrus. The entire story is so miraculous that it is hard to believe; nevertheless it's true.[7]

Jägerskiöld hints at several things in his letter. Apparently his main concern was the abuse of the walrus. Jägerskiöld was disgusted by the cruelty inflicted upon the animal, but as a scientist and head of the zoological collections, he could not conceal that he was content that he had acquired a walrus for his museum. Of utmost importance was the location *where* the walrus had been collected for the museum; that is, in Sweden. Even if the animal had come from a distant and foreign place, it was its final stop on the skerry outside Gothenburg that qualified it as Swedish fauna. This explains why the walrus is not inscribed in the register of foreign animals. The walrus was an addition to the plethora of wild Swedish animals and would enhance the fame of the museum displays. Jägerskiöld is also telling Jensen that the fact that the walrus had been on shore at Skagen would have consequences for the classification of Danish fauna. The walrus was a feather in Jensen's cap as well, "because the animal after its visit to northern Jutland also must be included to the fauna of Denmark".[8]

The cruelty shown towards the walrus was reported in Danish newspapers, and, as Jensen wrote, the act "caused quite a stir and indignation in wide circles".[9] In response to an inquiry initiated by The Danish Natural History Museum to verify that a walrus really had been seen in Skagen, Jensen received a letter from lighthouse assistant, V. Christensen.[10] Christensen confirmed that the newspapers gave a detailed description of the drama. At 10 p.m. on 5 January, the walrus came ashore between Skagen Lighthouse and Grenen, the northernmost point of Jutland. The man on duty at the signal station observed the beast through his binoculars and would later confirm that it was a walrus. The walrus was also observed by machine operator Valdemar Mariager, who shot at it with a rifle. Mariager had been quite close to the animal, and could confirm that it was a walrus. He had seen the animal's tusks at close range, and noticed that when it made its way back to the water, it used its tusks to haul itself forward. A second man had also seen the animal and agreed that it was a walrus. The wounded walrus lost a lot of blood while ashore, and continued bleeding before it reached deep water and disappeared to the south–east. The animal's motion, as described by Christensen, is typical of the walrus and is reflected in its generic name, *Odobenus*, which means "he who walks (*baino*) with his teeth (*odo*)" or "toothwalker" (Maxwell 1967: 72, King 1983: 69).

The Danish Society for the Prevention of Cruelty to Animals, Dyrenes Beskyttelse, published two articles containing further details about the walrus's visit at Skagen. J. Thamdrup, editor of *Dyrevennen (The Animals' Friend)*, condemned Denmark's reception of the walrus, noting that it reflected a typically negative attitude towards nonhuman animals: "'Rare guest is a welcome guest' one of our old sayings reminds us; but when it comes to such rare guests from the animal world, they are often 'welcomed' in a hideous, violent, and heartless way"(Thamdrup 1927a: 7).[11] Thamdrup described how Mariager, after having discovered the big animal, yet without having identified its species, ran back to the village to fetch a rifle. Mariager had then returned to the shore with both a rifle and a pistol, and accompanied by carpenter Ole Hansen, who was equipped with a sporting rifle: "It did not occur to any of them that a good camera, eventually a photographer here, would have been a more decent equipment with which to welcome the strange animal" (Thamdrup 1927a: 7).[12]

The animal arose when approached by the two men, and they could see that it was indeed a walrus. They then began to shoot: at a distance of four or five metres, Mariager fired his rifle at the animal's head, and the walrus began to move towards the water. Mariager then fired four shots at the animal's head from a close range of about one and a half metres. Finally he fired one shot from his pistol at very close range into the walrus's body behind its left flipper, and three shots to its head. Simultaneously, Ole Hansen fired his shotgun five times into the walrus's head. Blinded and bleeding, "it plunged into the water to swim away and later to receive the *coup de grâce* in the Swedish skerries" (Thamdrup 1927a: 7).[13] Mariager and Hansen were fined 20 Danish crowns each for their treatment of the walrus, and one of them had to pay an additional 30 crowns for hunting without a licence (Thamdrup 1927b: 42). The differing severity of the fines suggests that unlicensed hunting was deemed a more serious crime than animal torture.

Jensen received more sad details of the walrus's physical condition in Jägerskiöld's next letter. Taxidermist Skoog and his assistant David Sjölander, tough men accustomed to hunting, flaying, and dissecting dead animals, had been on the verge of tears when plucking shotgun pellets and wads out of the walrus's lips. There were shotgun pellets even in its hind flippers.[14] The eyes were smashed by shotgun pellets, "but whether this had happened in Skagen or in another place, the walrus did not have a chance to tell!" Jägerskiöld proclaimed ironically.[15] The dissection proved that the animal was a very old bull, "maybe sick", and lean.[16] Probably the walrus had lost its sight and had been too enfeebled to forage after it had been wounded, because the intestines appeared to be empty.[17] In a letter to Russel, written a year after the walrus had been killed, Jägerskiöld explained that the animal even had a wound behind the right eye, caused by a shotgun, and that the entire head had been inflamed.[18] Utter-Anders had indeed given the walrus a *coup de grâce*.

The abuse of the walrus made headlines in Swedish as well as in Danish newspapers, forcing Jensen to separate facts from rumours and the Danish Society for the Prevention of Cruelty to Animals to incriminate the guilty men. A Gothenburg newspaper reported that Jägerskiöld intended to publish a strong statement in a Danish newspaper about the abuse the animal had received during its time in Denmark.[19] But this was only a rumor; Jägerskiöld never made any statement to the Danish press, although the press did reflect the public's strong engagement with the walrus's sad destiny.

"... so miraculous that it is hard to believe"

The words are those of Professor Jensen: "You are right: the whole story is so miraculous, that it's hard to believe it's true".[20] The above reconstruction of the bull walrus's amazing journey from Greenland to Sweden, via Scotland, Norway, Holland, and Denmark, was made by Jensen, the same year the walrus had been shot on Swedish land (Jensen 1927a and b). As head of the vertebrate collections and chair of the Museum Board, Jensen corresponded with his colleague Leonard

Jägerskiöld, mostly about fish.[21] Jensen, an expert on Greenlandic fish species, and the zoology of Greenland and the Faroe Islands, compared the death of the beached walrus to contemporary reports about walruses in Scottish, Norwegian, Dutch, and Danish waters, and published his findings in two short articles in 1927. Jensen's aim was to reconstruct the animal's itinerary, from its starting point to its place of death. He was convinced that the walrus had set out from Danish East Greenland, and not from Norwegian Svalbard. Could there have been any political considerations in defining the walrus as Greenlandic and thus as Danish rather than Norwegian? Probably not: there are no allusions in Jensen's letters to the serious dispute over Greenland developing between Denmark and Norway.[22] The walrus had indeed been swimming in waters upset by political turmoil. In 1921 the Danish government had declared Greenland and its nearby waters to be under Danish jurisdiction, a decision that was strongly opposed by the Norwegian government, because it was unfavourable to Norwegian scientific exploration as well as to fishing and hunting in East Greenland.[23]

While the old bull walrus's odyssey was a miracle, its dead body was an unexpected contribution to the museum's collection of a nearly extinct animal. In 1927, after three centuries of intensive hunting, the Atlantic walrus population was close to extinction and "had become extremely rare" (Maxwell 1967: 69). A short communication in 1927 about the walrus that had been killed in Bohuslän emphasized the rarity of the animal: "… due to the hunters' war of extinction, these animals are even in their habitat decimated to such a high degree that an incident like this is even more remarkable".[24] The walrus has been hunted for its blubber, skin, and precious teeth. Walrus tusks are ivory, and continue to grow throughout the animal's life. Arctic people have used walrus tusks as raw material for fine objects, and the tusks have been traded as ivory. Walrus ivory does not turn yellow as quickly as does elephant ivory, and was for a time the preferred material for false teeth (Allen 1928 cited in King 1983: 69). The walrus's penis bone has also been historically sought after, and being up to 60 cm long, it has been something of an unusual collector's item.

Museum Item

The example, an old male, scarred so it almost looks tattooed on the entire enormous neck and the upper part of the body, souvenirs from fights for the females during the mating period – is bound for the Natural History Museum and will when time comes adorn its collections.[25]

The museum paid Anders Olsson 200 Swedish crowns for the walrus body; a rare and highly sought after museum item acquired for a relatively small sum. In total, the museum spent around 8,800 crowns on the acquisition of specimens in 1927, 1,000 crowns less than had been spent on the Ward gorilla acquired by Jägerskiöld twenty years earlier.[26]

The bargain initiated the animal's transformation from a dead body into a museum specimen. Leonard Jägerskiöld and taxidermist Hilmer Skoog left Gothenburg in the morning the day after the walrus had been killed. They stayed at Rörö overnight before returning to the museum. Skoog's short notes in his journal tell us which parts of the animal were desirable to keep for the museum: "Jan. 10–11 travelled to Rörö to take care of Walrus (sic) flayed and skeletonized the larger limbs of the same and taken specimens of tongue, throat, heart, stomach, spleen, caecum, aorta and urogenitalia."[27] In the general catalogue the list varies slightly, saying that an old bull's hide and skeleton, throat, heart, urogenitalia, stomach, spleen, caecum, a piece of aorta and cortex had been collected. While the hide has been mounted and displayed, the skeleton has never been mounted. Utter-Anders had fired a third shot through the animal's eye to ensure it was dead; the damaged skull is proof of this.

The next step in the preparation of the walrus was to make plaster casts of the flippers and the skull. Between 12 and 13 January Skoog made plaster casts of the head, neck, and flippers, and moulded them in February. A second task to be done immediately after the walrus had been taken into the museum was to separate the skin from the blubber and shave it. No flesh or blubber had been retained; a third of a walrus's total weight is blubber (Maxwell 1967: 68). The skin was then soaked in a salt solution. At the end of January and through February, Skoog also worked on other animals. A multitude of creatures from all over the world passed through the taxidermist's hand: he made a colour sketch of the beak and feet of an imperial penguin, plinths for a giant armadillo and a sun bear, completed a painting of a sea tortoise, and arranged the armadillo and the bear in the case displaying objects new to the museum's collection.

In March, Skoog worked on the tusks. The originals were to be kept with the skeleton, while models for their substitutes were moulded in plaster. The plaster models and an indication of the colour of the originals, were sent to Oscar Gottlow's "Celloid- & Konsthornvaru-fabrik", a firm that fabricated items of celluloid and artificial bone, situated in the Scandian town Eslöv. Gottlow reluctantly took up the intricate task of making copies of the tusks for the mount. Gottlow wrote to Skoog at the end of the month:

To imitate exactly the same colour is difficult, not to say impossible. I have contacted my supplier of raw materials abroad, and this firm has indicated that the colour that will be closest is no. 103, of which a small sample is included.[28]

A week later, Gottlow sent his estimate to the museum, offering to deliver two tusks made of synthetic resin, painted and ground, but not polished, for a sum of 50 Swedish crowns.[29] Skoog promptly ordered the expensive tusks the next day. Obviously the museum was short of money: The tusks and bill were delivered at the end of June. In early August, Gottlow sent the museum a reminder, and subsequently was asked to extend the date of due payment to the end of August. The museum did not pay before the middle of September and only after Gottlow had sent them a second reminder. However, he still signed his letter "I remain Sir, your obedient servant".[30]

Good taxidermy requires accurate vision. The preserved walrus reminds us of the importance of correctly emulated teeth, especially on an animal whose visual and cultural identity, as well as scientific

Fig. 33. The left and undamaged side of the cranium.

GNM_1042_1.

binomial, refer to teeth. Alas, Gottlow did not produce the correct colour; the tusks are too yellow to imitate the animal's natural teeth.

While Skoog was busy working on the mount, Jägerskiöld received letters from naturalists and various mercantile firms asking for pictures of the stuffed walrus. As discussed previously, photographs were important mediators of museum animals in a number of ways. The museum procured photographs to be used in its exhibitions, while producing its own photographs. In the case of the walrus, a photograph of a good fully mounted specimen appeared to be just as instructive as a photograph of its living counterpart, and in accordance with the aim of taxidermy: to substitute the live animal with a life-like specimen. And unlike photographers of live animals, the photographer was in this case in full control of his subject.

Jensen was the first to ask for a photograph of the walrus, in early March, "because, the animal after its visit to Northern Jutland must also be included in Danish fauna".[31] He was informed that the animal was still in the taxidermy room, but inquiries were made if he would instead like to have a plaster model sized 1:10, which Skoog was about to complete,[32] in addition to a photo when the walrus was finished. Jensen wanted both.[33] In April, he wrote that because the model of the walrus was to be put on display, maybe it would look best if patinated? He added: "Apropos the walrus once more! Would it be possible for you to let us have a photo of the cranium, of course we will cover the total expenses."[34] The way the two gentlemen expressed themselves in their correspondence, reflects their physical stature in an amusing way. Physically Jägerskiöld was a tall and stout man, while Jensen was the opposite. Jensen adorns his prose with polite phrases, while Jägerskiöld is straightforward and outspoken. Jensen would have to wait; there was still work to be done:

Dear friend! The cranium is being cleansed. My amanuensis refuses to photograph a dirty cranium, and because the matter doesn't have any haste, I will not use dictatorial command. This means that you will get photo when the cranium is ready and by then the plaster model should be dry and patinated in the most natural colour.

With my best greetings, yours sincerely, Jägerskiöld[35]

Fig. 34. The walrus observed and photographed by Redeke.
Skoog and Sjölander made their identification
partly based in this unfocused photograph.

The Dutch zoologist H.R. Redeke had witnessed and photographed the walrus when the animal passed by Helder on 11 November. Redeke had immediately published his observation in Holland, marking the first observation of a walrus in that country, and he was now curious to know if the walrus he had seen was the individual held in Gothenburg. Again, photographs proved important here. Redeke acquired a photo of the cranium with tusks, and asked for the animal's measurements.[36] He sent a copy of his photograph of the walrus to Jensen, who forwarded it to Jägerskiöld, asking Jägerskiöld if he thought it could be the walrus

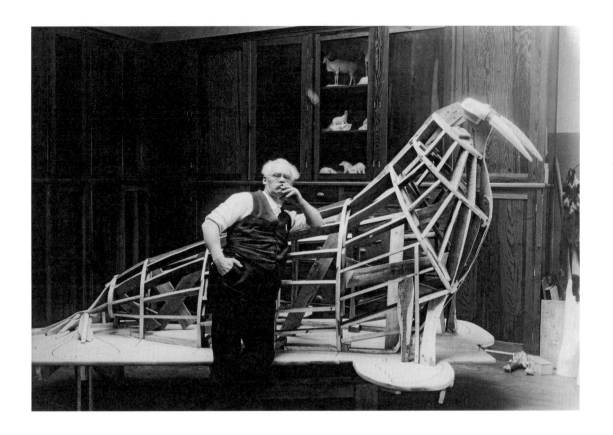

Fig. 35. Hilmer Skoog relaxes after having completed the work with the walrus template, the "stommen" or "trunk" on which to model the manikin. The date is 14 June, 1927. Behind him is a vitrine with small animal models made by Skoog. The models served as preliminary studies for the mount.

PHOTO: PAUL HENRICI. GNM_1156.

Fig. 36. Model, pre-study for taxidermy.

GNM_1042_1.

Redeke had observed. Redeke had estimated the length to be about 2.5 to 3 meters; the walrus in Gothenburg measured 3.32 meters from "snout to the tip of the tail". Skoog and Sjölander, being the only people to have seen the walrus before he was flayed, were convinced that the walrus seen and photographed by Redeke and the walrus under preparation was the same individual. Their identification was, among

CHAPTER 3

other details not mentioned, based on the skin pleats on the neck: the Gothenburg walrus had made a visit to Holland.[37]

The walrus body was first materialized as a small plaster model.[38] While Jensen considered a plaster model a useful didactic object, Director A. Jacobi at Museen für Tierkunde und Völkerkunde in Dresden, the Museums for Zoology and Ethnography, only displayed models of animals he could not display naturalized, such as whales. Jacobi was looking for the real thing: "I have for a long time been looking for a walrus hide for sale, as soon as I will succeed in this, I would like to use the Gothenburg specimen as a model".[39] Today the neat collection of sculptured animal models made by the taxidermists Hilmer Skoog and Björn Wennerberg as pre-studies for taxidermy, is kept in a glass case in the lunchroom, and is not available to the public.

In June, Skoog built the wooden skeleton on which the body would be modelled. He was photographed leaning on the skeleton while smoking, looking content and relaxed after having given the walrus its initial rough shape. At this point, Skoog was ready for a break. The photograph may have been taken on 14 June, the day he wrote in his journal: "Completed the construction of the skeleton for the walrus". This was to be his final entry: the following week, Skoog went on holiday, and on 15 July he died suddenly of a stroke, aged 57. Jägerskiöld had been reposing at Fagervik estate in Finland from the middle of June, but apparently he worried about the walrus. In a letter of 2 July Jägerskiöld asked his assistant, Sune Swärd: "How far did Skoog proceed with the walrus?"[40] And learned that "the walrus is in the same state as when the Professor left".[41]

Jägerskiöld must have been saddened by the sudden loss of such a skilled taxidermist, a hunting companion and friend: "A wreath with a blue and white ribbon 'To Hilmer Skoog from the Gothenburg Museum' would be placed on Skoog's bier".[42] But Jägerskiöld's letters reveal a persistently pragmatic attitude: how to complete the taxidermy of the walrus? The walrus had evoked expectations. Not only did Jensen and Redeke wait for photographs of the mounted animal, A. Gross Illustrations-Verlag in Berlin had in the middle of January asked for a photo of the walrus to be published in various illustrated magazines as "the giant walrus, the first to find its way to Denmark after the Ice Age".[43] Gross had wanted to borrow a photograph, and promised to return it in good condition. When they learned from Jägerskiöld that this was impossible because the walrus first had to be mounted, they asked for a picture of the professor with the walrus under preparation.[44] Another photo-bureau, Atlantic Photo-co in Berlin, contacted the museum in early June for a photo to distribute to newspapers and journals.[45] In short, the walrus's fame was increasing abroad, and the taxidermy work had to be completed as soon as possible.

Three days after Skoog's death, Jägerskiöld wrote to 'landshövding' Oscar von Sydow seeking permission to engage David Sjölander to complete the walrus and attend to other important tasks. Sjölander had been assisting Skoog in his work with the giant: "Momentous works – among these the huge walrus – are in progress. If these were not to be accomplished the museum will suffer irreparable damage".[46] Von Sydow gave his consent.[47] Sjölander modelled the walrus body and arranged the skin into a 'praktpjäs' – a showpiece.

Jägerskiöld referred to the 'praktpjäs' in a letter to the museum board, pledging them to promote Sjölander to the position of museum taxidermist.[48] It is clear that the walrus, after Skoog's sudden death, facilitated Sjölander's entry to a permanent position in the museum.

Jensen had reconstructed the animal's journey from Greenland to Gothenburg, including its exceptional visits to five countries, "the fabulous animal's vagabonding" as he described it.[49] He had published the route in both Danish and English. In January 1928, Jägerskiöld received a letter from George W. Russel. Russel asked for information about the Gothenburg walrus, especially about its tusks and its head: "… what I would particularly like to know is – Were the two tusks of your walrus of equal length or was one longer than the other? And what, if any, wounds showed about the head of the animal apart from that inflicted when it was killed?"[50]

The walrus Russel and his friends had followed at close range late in November 1926, had been seen in the same waters as late as in January, so the one observed at the same time off the Norwegian west coast had to have been another animal. The Shetland walrus had also "a raw red wound right in the middle of its forehead," identified by Russel as "undoubtedly a shot wound and if not a bullet wound must have been caused by shot discharged at very close range. It might I imagine have been caused by the Customs Officer at Hillswick who discharged a shot at the walrus from a few feet off".[51] In Russel's opinion, the odyssey of the walrus was that of three walruses: one with a splintered tusk observed in several places off the west coast of Scotland, a second with a fresh bullet wound in its head swimming in the Shetland archipelago for three months, and a third one crossing the North Sea with Rörö as its final destination.[52]

On Russel's last letter of 21 April, 1928, is written "intet svar" and "fått årsskriften" – "no answer" and "received the annual report". For some reason, Jägerskiöld chose to close correspondence with Russel. Was this because the Scotsman had deprived the walrus of some of its allure, or rather because male walruses, and their occasional vagabonding, did not warrant Jägerskiöld's attention once the walrus had been stabilized and inserted into the order of the museum? The most probable reason for not following up on Russel's rather convincing argument that the Shetland walrus was around for so long that it had to be a second animal, was that Jägerskiöld chose to be loyal to his colleague in the field, Jensen. In the annual report of 1927, Jägerskiöld mentions that Russel had reported one or several walruses seen off the Shetland Islands,[53] but omits Russel's clearly stated conclusion: "My own provisional summing up is that there was one walrus in Shetland, one in the west of Scotland, and a third across the North Sea."[54]. Jägerskiöld was acquainted with Russel's article in *Country Life*. But he was most likely unaware of that Russel was "an extraordinary well informed ornithologist with an incomparable knowledge of Shetland birds",[55] "the foremost authority in Scotland" on the birdlife of Shetland.[56] As an ornithologist he was well trained in observing wildlife, as a solicitor he was remembered as man who combined "intense local knowledge with a clear-thinking, logical mind".[57] To Jägerskiöld, Russel was after all an amateur and could not threaten Jensen's authority on the reconstruction of the walrus' journey, which was repeated in the museum guide of 1964.[58]

*Fig. 37. Walrus and Jägerskiöld in the Mammal Hall photographed
by the newspaper* Handelstidningen *in 1933.
Note the parallel curves of the two bodies.*

GNM_3038.

173

Even if Scotland should be omitted from his journey, the walrus remains a remarkable animal. By leaving Greenland and simply moving his body, the walrus had aroused the scientists Jägerskiöld, Jensen and Redeke, museum directors, German advertising agencies, taxidermists Skoog and Sjölander, the illegal hunters Mariager and Hansen, the heroic hunter Utter-Anders, animal protectionists, fishermen in various European communities, and journalists in Denmark and Sweden.

Back to Nature – Fading Fame

The walrus's arrival in Swedish waters was an extraordinary event that evoked considerable curiosity for a limited period. In the annual report of 1927, Jägerskiöld described the walrus as "the most noteworthy" of the museum's acquisitions that year, and as a "valuable attainment".[59] Nearly one hundred years after it was stranded on the small skerry outside Rörö, the animal's remarkable journey from Greenland to Sweden has been largely forgotten. Today it is remembered as "Valrossen från Rörön" – "The Walrus from Rörön". The forgetting of the walrus's odyssey must also have been effected by the relocation of its stuffed body within the museum.

When put on display in March 1928, the walrus was given a central place in the Mammal Hall, on a square, low, wooden plinth that had previously supported a walrus skeleton. Visitors could walk around the animal and study the body from three sides. There was no glass vitrine to prevent the most curious from feeling the quality of the fur and the ruggedness of the scars. The display underscored the animal's

Fig. 38. The entrance to the Mammal Hall.
Note the symmetry as a principle in the display. 1934.

PHOTO: DAVID SJÖLANDER. GNM_3369_1.

singularity, as well as its dislocation from the natural world. What the audience saw was an individual walrus, just as detached from its origins as the other mounted animals surrounding it; yet its disconnection within the Mammal Hall also pointed to its final, fatal manoeuvres into the foreign waters of Sweden. Exotic cats were watching the walrus from a large vitrine on the right hand side. Three seals overlooked it from behind, while on its left a selection of herbivores in their cases stared at the walrus with big, soft, glass eyes. The walrus itself, with its small head proudly held high, was turned towards the main entrance of the hall. The immense animal, once weighing two tons, towered over two small seals displayed in front of it. The walrus was the first animal to catch the visitors' eyes as they entered the Mammal Hall. Like most large mammals in central locations within museums, the display demonstrated both the grandeur of the species and the mount's importance as a museum showpiece.

The prominent positioning of the walrus began to lose its stability when David Sjölander initiated his work on the African elephant, first in the lecture room, and later in the Mammal Hall. Several animals had to be moved to make space for the elephant – among these was the walrus, which was moved closer to the entrance and to the right, within a freestanding glass case. People could still study the animal from all angles, but the glass prevented a close examination of the body's peculiarities. Perhaps most significantly, the walrus was no longer displayed in the centre of the hall – in 1952 this privileged position instead went to the elephant. Eventually, a huge hippopotamus squeezed the walrus out of the Mammal Hall altogether. Taxidermist Björn Wennerberg, who had shot the animal at Lake Ellen in Ethiopia in 1958, completed the hippo mount in 1960, and it was displayed in the walrus's former place on the first of September of that year.[60] Together with numerous species from Africa, the African big mammals today dominate the Mammal Hall. The walrus was relocated to the Whale Hall to join the seal species. Here it was relegated to a corner under a balcony.

Fig. 39. *The walrus is transported out of the Mammal Hall.*

PHOTO: BJÖRN WENNERBERG. GNM_6964.

The walrus's transferral to the Whale Hall was more than a mere change of spatial location – it changed the animal's meaning as well. From being a prominent centrepiece in the Mammal Hall, the old bull was now confined – and almost hidden – in a corner of the densely populated Whale Hall. A bull walrus is a huge animal, but it is difficult to compete with a blue whale – even a blue whale calf. The size of the famous Malmska Valen – Malm's whale – dwarfed the walrus. The walrus became utterly reduced by the new installation made with and for the specimen. It was now locked in a diorama box, visible only through a small window placed so high upon the wall that children must climb a ladder in order to look through it. The box is narrow and the space at its front so restricted that it is impossible to see the entire body of the animal in full view. The walrus is displayed in profile from the right. It is a mild irony that the head is seen from this side when we know that Utter-Anders fired the last shot through the animal's right eye and smashed the cranium on this side. His nose is invisible, and only a few of the impressive whiskers, so essential for feeding on clams, can be seen.[61] The body is placed against a painted landscape of East Greenland, rendered by the Swedish artist Allan Andersson. It is lying on a rock with drifting ice and snowy mountains in the background. The scene's light indicates Arctic spring or early summer, although the walrus's biography is linked to autumn and winter.

In her famous analysis of the African Hall in the American Museum of Natural History, Donna Haraway states: "A diorama is eminently a story, a part of natural history. The story is told in the pages of nature, read by the naked eye" (Haraway 1989: 29). Dioramas intend to anchor the animals in their natural habitats. Karen Wonders claims that the diorama reflects "man's aesthetic delight in the wonderful forms and colours of animate beings and the intricate temporal patterns or configurations of their life stories, behaviour and ecological settings" (Wonders 1993: 221). Surely Wonders refers to the animals' *natural* life stories, that is, those not intersected by human beings. The problem, of course, is that all mounted animals in natural history museums are there because of humans. Therefore, the habitat diorama deprives the animals of their individuality, and of their biography, beyond the deprivation of the unadorned vitrine.

Fig. 40. The walrus in the diorama box.

PHOTO: ANDERS LARSSON, GNM.

The naturalism inherent to the habitat diorama also works to obscure the potentially creepy sensation of looking at remains of dead animals. Modern zoos are constructed to prevent people from asking unpleasant questions about the animals' confinement; the habitat diorama invites the audience into an idealised, reconstructed natural scene, and renders the animal's death invisible. By confining the walrus to a corner under the balcony, against a backdrop of Greenlandic nature, the museum has converted an individual animal into a specimen of *Rosmarus odobenus odobenus*. The historic individual is displayed as a bull walrus in East Greenland, thus disrupting its connection to local history. A small sign reads that the walrus was shot by Anders Olsson at Rörö in the skerries of Gothenburg, but provides no explanation of how this singular walrus connects Greenland and Rörö.

The mounted walrus is now close to a hundred years old. The right flipper is broken, and the skin is cracking. The artificial teeth have yellowed over time and are now closer to a shade of orange. But the mount is masterfully done and presents a monument of a species, one that was close to extinction when this blind and suffering representative climbed a small skerry at Rörö in 1927. Perhaps the time has come to open the box, so that visitors can see this splendid piece of taxidermy, learn about the animal's odyssey, and appreciate its individuality.

The Ugly Walrus

The most thought-provoking lesson to be learned from the walrus's biography is human ignorance and cruelty. Bullets followed the walrus wherever it appeared in the northern seas. Russel provides a detailed description of the deep, fresh wound in the Shetland walrus's head: "The walrus we saw here had a raw red wound right in the middle of its forehead, I should say just at the place where on a cow a butcher strikes his pole axe. It would have been I should think about an inch and a quarter to an inch and a half in diameter and was deep and well defined, such a wound would not heal up without the lapse of some time."[62] The Gothenburg walrus died with an inflamed head, smashed eyes, and flippers filled with shotgun pellets. Why did the walruses so strongly fuel the desire for a trophy, and to so little a degree, awe for the unfamiliar? Thamdrup explained the cruelty inflicted upon the walrus simply in terms of tradition: rare and big guests from the animal world have historically been welcomed with hostility and death. A strange dead animal is a trophy, a materialized proof of a deed, an annihilation of an unknown threat.

To this should be added comments about the aesthetics involved: as opposed to the baby seal with its expressive eyes, the walrus doesn't exhibit the features of a "cute" animal. Obviously, animals considered ugly are more vulnerable to being killed by humans than are their cute counterparts. A threatened species, lacking charisma, has less chance of gaining the public attention and affection it needs to survive. Perceptions of animal cuteness are transferred from human to non-human beings. We are inclined to feel affinity with creatures that remind us of the human child's physical appearance, creatures that conform to what Konrad Lorenz called the *Kindchenschema*, "the infant schema for the aesthetic proportions of the heads of human and non-human animals considered to be cute" (Genesko 2005: 3). The walrus does not exhibit these proportions, having neither a big head relative to its body, a protruding forehead, large and low-lying eyes, nor podgy cheeks (Lorenz 1971: 154–155). The walrus's body is virtually the opposite: the head is extremely small compared to the oversized body, so small that the head almost drowns in the enormous neck, its eyes are small, its whiskers recall an old man, and the tusks add a final oddity to this generally bizarre body.

History repeats itself. On New Year's Eve 2012, a young bull walrus crept ashore on a floating bridge in Kristiansund, a city on the northwest coast of Norway, and was shot. The walrus is a totally protected species in Norway. Ironically, the unlucky walrus was shot as part of an initiative of the local division of Mattilsynet, the Norwegian Food Safety Authority. Mattilsynet is also responsible for animal welfare in Norway, and disseminates information related to the Animal Welfare Act. The kill provoked heavy protests, and walrus experts claimed, contrary to the representatives of Mattilsynet and the local Game Committee, that the animal was tired after a long swim, but looked perfectly healthy. After the killing, another peculiar thing happened: the carcass, instead of being delivered to the authorities, was transported into deep water and dumped. Norges Miljøvernforbund, Green Warriors of Norway, reported the Norwegian Food Safety Authority to the police for fauna crime. The police have since dismissed the case.

Notes

1 http://www.marinemammalscience.org/index. php?option=com_content&view=article&id=512& Itemid=320. Retrieved 17.09.2013.

2 Oral information from wild life artist and ornithologist Viggo Ree. There are some 31 records of walruses on the Norwegian coast between 1900 and 1967 (Brun *et al.*, 1968 cited in King 1983: 67).

3 Årsberättelse för Naturhistoriska museet i Slottsskogen 1928. *Göteborgs Museums Årstryck* 1929, 11.

4 Gothenburg Natural History Museum introduced in 1911 leg ring marking of birds as a scientific method in Sweden (Mathiasson 1983: 50).

5 Letter from V. Christensen to A.S. Jensen 11.01.1927; letter from A.S. Jensen, to L.A. Jägerskiöld 17.01.1927; letter from A.S. Jensen to L.A. Jägerskiöld 25.01.1927.

6 *Torpedo marmorata* is a fish, the marbled electric ray.

7 "Käre kollega.
Tack för upplysningen om *Torpedo marmorata*.
Antagligen har det stått i danska tidningar om valrossen, som besökte skagen (sic) 7 el. 8 ds. De spottade honom full med hagel, förladdningar m.m. i synen och litet varstans. Tacka phan (sic) för att det plågade och blinda djuret gav sig iväg. Det underbara var emellertid, att han for över till Bohuslän och lade sig på ett skär några 100 m. från vår skickligaste sälskytts stugu på Rörö. Det var den 9/1. Mannen som annars är ute på sillfiske var hemme, då det ju var söndag, och av hans mauser fick valrossen en snabb och smärtfri död. En gammal fullvuxen hane med kraftiga betar tillhör numera Göteborgs Museum. Ni kan alltså lita på tidningarnas uppgifter om valrossen förekomst på Skagen, vilket ju angår Eder fauna. Kunde Ni prygla eller åtminstone brännmärka de uslingar, som på detta sätt behandlade valrossen ifråga, skulle det glädja mig. Hela historien är så underbar att man knappast vill troden (sic); icke förthy är den sann".
Letter from L.A. Jägerskiöld, Gothenburg, to A.S. Jensen, Copenhagen, 14.01.1927. Gothenburg Natural History Museum's copybook 23.02.1926–6.06.1927.

8 "... da Dyret efter dets Besøg ved Nordjylland jo ogsaa maa regnes med til Danmarks Fauna". Letter from A S. Jensen to L.A. Jägerskiöld 07.03.1927.

9 "... har vakt Opsigt og Forbitrelse i vide Kredse". Letter from A.S. Jensen to L.A. Jägerskiöld, 17.01.1927.

10 Letter from V. Christensen to A.S. Jensen 11.01.1927.

11 "'Sjælden Gæst er velset Gæst' siger et af vore gamle Mundheld; men med Hensyn til saadanne sjældne Gæster fra Dyreverdenen, saa er det ofte en hæslig, raa og hjerteløs Maade, de bliver 'velset' paa" (Thamdrup 1927a: 7).

12 "Det faldt vist ingen af dem ind, at et godt Fotografiapparat, eventuelt en Fotograf her havde været en værdigere Udrustning at modtage det fremmede Dyr med" (Thamdrup 1927a: 7).

13 "... inden det blindet og blødende naaede at kaste sig i Havet og svømme bort for siden at faa Naadeskuddet i den svenske Skærgaard" (Thamdrup 1927a: 7).

14 Letter from L.A. Jägerskiöld to A.S. Jensen 19.01.1927.

15 "Ang. blindheten voro ögonen söndermosede men ej variga. Skadan kan alltså ej vara mycket gammal. Vi äro övertygade om att den beror på hagelskott. Men om det skett vid Skagen el på annat ställe

hann ej valrossen att tala om!".
Letter from L.A. Jägerskiöld to A.S. Jensen
19.01.1927.
Letter from L. A. Jägerskiöld to A..S. Jensen
12.03.1927.

16 Letter from L. A. Jägerskiöld to A.S. Jensen
09.03.1927.

17 *Fauna och Flora* 1927, 1, 43.

18 Letter from L.A. Jägerskiöld to G.W. Russel
11.01.1928.

19 Letter from L.A. Jägerskiöld to A.S. Jensen
29.01.1927; "Rörö-valrossen orsakar rabalder i
Köpenhamn", *Handelstidningen* 18.01.1927;
"Affæren faar Efterspil i Retten", *Berlingske
Tidende*, 18.01.1927.

20 "De har Ret: hele Historien er saa vidunderlig,
at man knap skulde tro, den var sand".
Letter from A.S. Jensen to L.A. Jägerskiöld
17.01.1927.

21 As an expert on Greenland's fish populations
A.S. Jensen contributed to develop its fishery
(Laursen 1954: 522).

22 A letter from Direktøren for Grønlands Styrelse
October 1932 indicates that Professor Jensen was
involved in the political crisis between Denmark
and Norway. Statens Naturhistoriske Museum,
T 788 Adolf Jensens embedsarkiv. Jensen wanted
to red-list the muskox in Greenland to stop
Norwegian exploitation of the species.

23 The dispute over Greenland culminated on 26
June, 1931, when five Norwegians occupied a
territory around the Norwegian radio station
in Myggbukta and named the territory Eirik
Raude's Land, after the tenth-century Norwegian
Viking explorer who founded the first settlement
in Greenland. The famous explorer, lawyer
and archaeologist Helge Ingstad was appointed

'sysselmann', sysselmann being the medieval
denomination of the king's highest official in a
county. The Norwegian government supported this
illegal action and undertook an official occupation
on 10 July. On 5 April, 1932, the dispute was
resolved in the Permanent Court of International
Justice in the Hague, when the court declared the
Norwegian claim to sovereignty to be invalid. For a
detailed account of the Norwegian interests at stake
in the so-called "Grønlandssaken", "the Greenland
case", see Drivenes 2004.

24 "... då dessa djur genom fångstmännens utrotnings-
krig i så hög grad decimerats även i deras egentliga
hemvist, är en tilldragelse som denna än mera
märkvärdig". En valross skjuten i Bohuslän. *Flora
och Fauna* 1927, 44.

25 "Exemplaret – en gammal hane, ärrig så det nästan
verker tatuering på hela den väldiga halsen och
framkroppen, minnen efter strider om honorna
under parningstiderna – skall til Naturhistoriska
museet och kommer i sinom tid att pryda dess
samlingar". "Valrossjakt i Göteborgs skärgård",
Göteborgs Handels- och Sjöfartstidning, 11.01.1927.

26 Berättelse rörande Göteborgs Musei Zoologiska
avdelning för år 1927, *Göteborgs Museums Årstryck*
1928, 13–14.

27 "Jan. 10-11 Utrest till Rörö for tillvaratagande av
Valross flått och råskeletterat samma och tagit
preparat av tunga luftstrupa hjärta mage mjälte,
blindtarm aårta och urogenitalia". Skoog's journal
10.01.1927, archive number 369.

28 "Att få en precis sådan färg är svårt för at icke säga
omöjligt. Jag har varit under tiden i förbindelse
med min utländske råvaruleverantör och påpekade
denna firma, att den färg som närmast skulle
komma i fråga vore färg n:r 103, varav ett litet prov
närslutes".

Letter from O. Gottlow to H. Skoog 30.03.1927.

29 Letter from O. Gottlow to H. Skoog 08.04.1927.

30 "Alltid /framgent/ till Eder tjänst, har jag äran
teckna Högaktningsfullt" .
Letters from O. Gottlow to Naturhistoriska Museet,
Gothenburg 09.08 and 16.09.1927.

31 "… da Dyret efter dets Besøk ved Nordjylland jo
ogsaa maa regnes med til Danmarks fauna".
Letter from A.S. Jensen to L.A. Jägerskiöld
07.03.1927.

32 Letter from L.A. Jägerskiöld to A. S. Jensen
09.03.1927.

33 Letter from A.S. Jensen to L.A. Jägerskiöld
11.03.1927.

34 "Á propos Hvalrossen endnu en Gang! De
skulde vel ikke ville lade os faa et Fotografi af
Kraniet, selvfølgelig mod Godtgørelse af alle
Omkostninger?".
Letter from A.S. Jensen to L.A. Jägerskiöld
04.04.1927.

35 "Käre vän.
Kraniet är under rengöring. Min amanuens
vägrar att fotografera ett smutsigt kranium och då
saken egentl. ej är brådskande vill jag ej använda
maktspråk! Ni får alltså fotografien, när kraniet blir
färdigt och då torde gipsmodellen vara torr och
patinerad i naturligaste färg.
Med de bästa hälsningar
Eder förbundne
Professor Dr. L.A. Jägerskiöld, GÖTEBORG".
Letter from L.A. Jägerskiöld to A.S. Jensen
13.04.1927.

36 Letters from H.R. Redeke to L.A. Jägerskiöld
29.07.1927 and 16.08.1927; letter from A.S. Jensen
to L.A. Jägerskiöld 07.03.1927.

37 Letter from L.A. Jägerskiöld to A.S. Jensen
09.03.1927.

38 Skoog's journal 16.03.1927, archive number 369.

39 "Ich suche schon seit längerer Zeit eine
Walrosshaut zu kaufen, sobald mir dies gelingt,
würde ich für die Montierung gern das Göteborger
Modell zum Muster nehmen ...".
Letter from A. Jacobi to L.A. Jägerskiöld 04.04.1927.

40 "Hur långt hann Skoog på valrossen?". Letter from
L.A. Jägerskiöld to S. Swärd 02.07.1927.

41 "… står på same sätt, som när Professorn reste".
Letter from S. Swärd to L.A. Jägerkiöld 30.06.1927.

42 Letter from L.A. Jägerskiöld to O. von Sydow
18.07.1927.

43 "… dem Riesen-Walross, dem ersten, das seit
der Eiszeit den weg nach Dänemark gefunden
hat". Letter from A. Gross Illustrations-Verlag
to die Direktion des Museums von Gotenburg,
Dänemark (sic) 17.01.1927.

44 Letter from A. Gross Illustrations-Verlag to
L.A. Jägerskiöld 24.01.1927.

45 Letter from Atlantic Photo-Co, Berlin to
L.A. Jägerskiöld, 03.06.1927.

46 "Men vi kunna omöjligen vara utan en man
tills Styrelsen i dessa frågor fattat beslut.
Maktpåliggande arbeten – bl.a. den stora valrossen
– äro i verket. Skulle de nu nedläggas lede muset
ohjälplig skada".
Letter from L.A. Jägerskiöld to O. von Sydow
18.07.1927.

47 Letter from O. Von Sydow to L.A. Jägerskiöld
01.08.1927.

48 Letter from L.A. Jägerskiöld to Styrelsen för
Göteborgs Museum 29.12.1927

49 "… det eventyrlige Dyrs Omstrejfen".
Letter from A. Jensen to L. Jägerskiöld 11.03.1927.

50 Letter from G.W. Russel to L.A.. Jägerskiöld
04.01.1928.

51 Letter from G.W. Russel to L.A. Jägerskiöld
    04.02.1928.

52 Letter from G.W. Russel to L.A. Jägerskiöld
    21.04.1928.

53 "Om detta dyrbara förvärv har inngått uppgifter,
    som visa djuret sannolikt i början av oktober 1926
    besökt Shetlandsöerna. Mr. G.W. Russel i Lerwick
    (Shetland) uppgiver att under hösten – vintern 1926
    en eller flera valrossar varit synliga i Shetland".
    Berättelse rörande Göteborgs Musei Zoologiska
    avdelning för år 1927, *Göteborgs Museums Årstryck
    1928*, 13.

54 Letter from G.W. Russel to L.A. Jägerskiöld
    21.04.1928.

55 "The Late Mr George W. Russel", *The Shetland
    News*, 18.08.1953.

56 "Death of former Official", *Shetland Times*,
    14.08.1953.

57 "The Late Mr George W. Russel", *The Shetland
    News*, 18.08.1953.

58 *Naturhistoriska Muséet Göteborg Vägledning 1964.*

59 "Märkligast är dock en valrosshanne", "Om detta
    dyrebara förvärv". Berättelse rörande Göteborgs
    Musei Zoologiska avdelning för år 1927, *Göteborgs
    Museums Årstryck 1928*, 13. Jägerskiöld reports
    incorrectly that the walrus was shot on 10 January
    instead of on 9 January.

60 Björn Wennerberg succeeded David Sjölander as
    museum taxidermist in 1953.

61 The walrus's preferred diet is benthic bivalve
    mollusks, especially clams. They locate the
    clams by sticking their sensitive whiskers into the
    sediment.

62 Letter from G.W. Russel to L.A. Jägerskiöld
    04.02.1928.

Loxodonta africana africana, male, 40 years+

Provenance: Huila province in Portuguese West Africa (Angola)

Date of death: December 4, 1948

Collected by: David Sjölander (1886–1954)

Mounted by: David Sjölander 1949-52

Owner: The Gothenburg Natural History Museum

Condition: Good

Entry in general register: 1949-8770

Entry in register of foreign mammals: Ma.ex. 1107

Entry in Collectio anatomica: 15.1315

Place in the museum: Body displayed in the Mammal Hall

Skeleton in the Bone Cellar

Red-listed: Vulnerable

Collected : The African Elephant   4

On 4 December, 1948, David Sjölander, taxidermist at the Gothenburg Natural History Museum, shot an old bull elephant in the Huila province of Portuguese West Africa, now Angola.[1] For some time he had studied the animal's habits and movements, and perhaps he had admired the grace of its huge body. Sjölander took advantage when the elephant paused for a midday nap under a shady tree, while waving his ears and occasionally sprinkling sand on his enormous body. Then he fired from a distance of about 10 meters. The bullet entered the body behind the right foreleg, penetrating the elephant's heart and bringing him to the ground with his legs folded underneath him. The elephant trumpeted a couple of times, tried to haul himself up by fastening his trunk around the tree, and died. To be safe, the locals fired an additional bullet into the elephant's hindquarters.[2]

Elephants have followed long, dramatic, spectacular, painful and lethal routes into menageries, zoos, and museum collections. Historically, individual elephants – the majority have been Indian elephants – have been protagonists of grand narratives, discussed, described, and depicted in their own time. Their remains, most often their skeletons, have been displayed in museums.[3] This chapter examines the movements and transformations embodied by an African bull elephant collected and mounted by taxidermist, photographer, and naturalist, David Sjölander. The elephant mount in the Gothenburg Natural History Museum is considered to be one of the best in the world, according to Jeanette Setterberg, former taxidermist at the Swedish Museum of Natural History. In her opinion, David Sjölander was the most gifted taxidermist to have worked in the Gothenburg Natural History Museum, and his works led the museum to be considered one of the most beautiful in Europe (Setterberg 1989: 27). The elephant mount became "the centrepiece of our displays", director of the museum and ichthyologist Orvar Nybelin wrote on the occasion of Sjölander's 65th birthday, naming him "an unrivalled master".[4]

The motives, efforts, and actions that led to the death of the old bull elephant, and its relocation from West Africa to Gothenburg, were deeply embedded in the collecting tradition and the professional climate of the natural history museum, and in the spirit of the taxidermist. A striking element of the dead *Loxodonta's* biography is the sheer physical labour involved in flaying, skinning, cooking, transporting, and mounting the animal itself. As I put together pieces of the elephant's biography, however, it became increasingly clear that I was also reconstructing a fragment of David Sjölander's *vita*. Why did Sjölander subject himself, at the age of 62, to the toil that appears to have ruined his health? The collecting and mounting of the elephant was to be his last and also his most strenuous work, which, if successful, would require a variety of skills, such as logistics, knowledge of how to travel, hunt, and collect in the wilderness, and an intimate experience of how to transform a dead animal body into a masterpiece of taxidermy.

Sjölander possessed these abilities. He had previously undertaken great journeys between 1919 and 1923, in China, Tibet, Mongolia, the Philippines, and Sumatra, partly collecting for the Swedish Museum of Natural History and partly shooting nature films for Svensk Film-industri, the Swedish Film Industry.[5] Yet there is an element of defeat in Sjölander's African expedition that puzzled me, as I gradually made my way through the story of the elephant's collection, a story which reads rather like an ageing man's efforts to dazzle his audience by the means and deeds of a bygone era.

There are few primary sources in the museum's archives to shed light on Sjölander's African expedition. Written and photographic material from the expedition has disappeared from the museum. The family has his expedition notebook and a collection of unsorted photographs; I was not able to borrow the notebook for perusal. Unfortunately, Sjölander removed documents and photos from the museum that could elucidate his work in general and the biography of the elephant in particular. He was an excellent photographer and had as a young man been employed at the famous Swedish photographic equipment company, Hasselblad. In the early 1920s he made nature films in China and Tibet, and for this achievement he was rewarded with a silver medal by the Swedish Academy of Science in 1924. The only two existing photographs from Sjölander's African expedition in the museum's archives were taken by Vasco Ferreira, a Portuguese hunter hired for the expedition.

One story making the rounds in the museum is that Sjölander put all the material inside the elephant; that the mounted elephant is a kind of sealed archive, a gigantic container for the elephant's and Sjölander's shared history. This story is strongly denied by others, however. Another possible explanation of the absent material is that Sjölander may have destroyed it. Luckily, five letters written by Sjölander to Director Orvar Nybelin from Portuguese West Africa have been saved. Minutes from the board's meeting, and letters from financial sponsors and donators of equipment, are kept in the archive, as well as Sjölander's list of collected specimens and many photographs documenting the mounting of the elephant. Three bills for the collection of the elephant also remain, as do many newspaper clippings. With this material, I have been able to trace some stages of the elephant's relocation from Portuguese West Africa to Sweden.

An Elephant Dream

The human protagonist of this mini-drama, David Sjölander, shared with the famous and historically renowned American Carl Akeley (24 years his senior) an obsession with achieving the perfect taxidermic mount. However, apart from being highly skilled taxidermists, and being socialised into the culture of the natural history museum as young men, their working conditions differed radically in many ways. Carl Akeley was employed by two of the most prestigious museums in the United States, the Field Museum in Chicago, and later, the American Museum of Natural History in New York. Akeley was given several opportunities to travel to Africa and hunt animals in order to realise his grand visions, first for the elephant display "Fighting African

Elephants" in the Field Museum, and later for the Hall of African Mammals in the American Museum of Natural History (although he died before it was completed). Sjölander, on the other hand, worked in a comparatively modest museum, mounted animal skins collected by others – a complicated task since he often had neither their measurements nor parts of their skeletons – and birds he collected himself every summer in Swedish Lapland. Their ability to influence the mounted animal displays also differed. While Akeley successfully recreated the appearance of African fauna in their natural habitats, Sjölander never designed a mammalian habitat diorama. Except for the magnificent African elephant that stands freely in the middle of the Mammal Hall in the Gothenburg Natural History Museum, Sjölander's mounts were put into the systematic displays.

The biography of the Angola elephant is intertwined with that of a tiny male elephant calf, displayed in a glass case vis-à-vis the huge bull. Sjölander mounted the hide of a one month old calf in 1948, just a few months before he left Gothenburg for Portuguese West Africa. The museum had bought the hide, skull, and leg bones of the baby elephant in 1912. The seller was Magnus Leyer, a Swede in British service in south-east Rhodesia, now Zambia. Correspondence with Leonard Jägerskiöld reveals that Leyer supplemented his income by selling skins, horns, and bones to museums and taxidermists.

As an elderly man, Leyer refused to be remembered as a big game hunter, if one trusts an interview published in the Gothenburg newspaper, *Handelstidningen*, printed on the occasion of Leyer's 65th birthday, under the headline "Elephant Trunk Tastes Good, to

Break Zebras Is Troublesome". Leyer explains that, according to white people, the only edible parts of an elephant are the trunk and the heart, whereas the natives cook even the skin for food; and he refers to his killing of 54 elephants merely as a kind of coincidence. According to the journalist, Leyer did not deny "that he had shot discernible big game in his days, and he verified the number of killed elephants, but … is very exact about pointing out that he had never been a professional hunter, and had really mostly by coincidence happened to shoot a lot of game during the long journeys he had undertaken in his vast county as a British official".[6]

The elephant calf is first mentioned in a letter dated 17 January, 1912. While positioned in a thick shrub, Leyer shot a female elephant, not noticing she had a calf:

I have now prepared, for musei purposes, a male Elephant calf, about a month old. Last December I happened to shoot an elephant in dense papyrus, and was sorry to find it was a cow with calf. I had not enough milk, so the calf died. Best wishes for the New Year! Yours sincerely M. Leyer.[7]

At this time, Leyer had sent "a large consignment, 13 bundles, of Game Heads and Skins", all collected and prepared by himself, to the Swedish Consul General in Cape Town, to be dispatched to Gothenburg.[8] When the carriers had been on their way to Cape Town for six weeks, Leyer composed a detailed list of the shipment's contents. The list exemplifies a big game hunter's knowledge of how to convert a killed animal into a valuable commodity. Skins had been treated with arsenic

Fig. 41. Elephant trophies. Abercorn Division, Tanganyika District, Rhodesia (Zambia), April 1910.

PHOTO: MAGNUS LEYER. GNM _5220_437.

soap, turpentine, paraffin, and naphthalene to prevent insect damage. Each specimen was identified by name, sex, and the place where it had been shot. Necessary measurements of the bodies had been taken. He also indicates where on the body the bullets had penetrated the hide.[9] Bullet holes may disrupt the completion of a naturalistic mount – they undermine the illusion of nature and hence are rarely to be seen on mounted animals. An eland buck shot by Leyer on 5 September, 1910, is described as a "youngish animal, of proper 'eland' colour i.e. not yet acquired the 'blue' coat of the very old bulls. 3 shot-holes 1. Side: shldr, lungs and thigh".[10]

The hide and antlers were used to recreate the animal as a trophy or a specimen. Any big game hunters who aimed at converting animals into money were obliged to master more than accurate marksmanship. Skinning and drying was "a delicate operation that sportsmen were advised to oversee personally" (Ritvo 1986: 252). Magnus Leyer treated his game as potential natural history specimens. He made the necessary arrangements to offer material of good quality, treating the hides chemically, and indicating the date and location of the "collecting", information that enhanced the scientific value of the specimen itself.

The remains of the baby elephant were sent from South Africa to Gothenburg in May 1912. In a letter containing accurately specified measures of the animal, Leyer underlines that the museum now will receive a unique and rare specimen:

I hope you will take this for the Museum, and you will be the best judge of a fair price for it. I might say that the capture of this young animal was entirely accidental, is not allowed by law, and therefore this is an almost unique occasion and opportunity to acquire such a specimen. In case you would like to take it over from me, I give below its measurements.[11]

Then Leyer lists 27 different measurements, and adds "if these measurements are not perfectly clear or sufficient, I can send a sketch

Fig. 42. Dead elephant calf 1912. The tape measure indicates that animal is destined for taxidermy. Photographer: Magnus Lejer.

with measurements noted on. I enclose herwith (sic) a photograph of the animal."[12]

Leyer's measurements proved to be sufficient when David Sjölander mounted the elephant calf's hide 36 years later. Sjölander must have been acquainted with the hide for many years, and appreciated

Fig. 43. The sheet with Leyer's measurements of the elephant calf.

PHOTO:

ANDERS LARSSON,

GNM.

I hope you will take this for the Museum , and you will be the best judge of a fair price for it . I might say that the capture of this young animal was entirely accidental, is not allowed by law, and therefore this is therefore an almost unique occasion and opportunity to acquire such a specimen .

In case you would like to take it over from me, I give below its measurements :-

Trunk, tip to root - 1' 10"
Root of trunk to between ears - 1' 2"
Ears to withers -   11"
Withers to croup - 1' 3"
Croup to root of tail - 1'
Length of tail - 1' 4"
Trunk, thickness at root - 9½"
      "   "   " end - 4½"
Ear, hor. length - 11½"
   ", vert. " - 1' 6"
Neck, circumference   - 3' 1"
     " depth   -   9"
Anterior girth, circ. 3' 7"; depth - 1' 2"
Middle   "   "   3' 7½" ; "   - 1' 3½"
Posterior "   "   4' 2" ; "   - 1' 4"
Foreleg , length top of shoulder to sole - 3'
Upper arm, circ. - 1'4" ; depth J- MM ,6" ,
Elbow,   "   - 1' 4" ; "   = 6"
Fore arm , "   - 11½" ; "   = 4"
Fore knee "   - 1' ;   "   - 4½"
Wrist   "   - 11" ; "   - 3½"
Fore foot, "   - 1'4½" ; long diam. - 6½", short diam. 5" .
Hind leg , height from croup - 3' .
   " knee, Circ. - 1'5½"; depth J- 7"
Shank ,   "   - 10" ;   "   - 3½".
Hind foot, circ. - 1'6" ; long diam. - 7½"n short diam.- 4" ;
Tail, thickness at root - 5½" .

If these measurements are not perfectly clear or sufficient, I can send a sketch with measurements noted on . I enclose herwith a photograph of the animal .

We are all well after our long journey from the neighbourhood og Lake Mweru to this Station, which is at the Luangwa River's confluence with the Zambesi . With kind regards from us both ,

                    Yours sincerely
                                    M. Leyer

Fig. 44. The mounted elephant calf.

PHOTO: ANDERS LARSSON, GNM.

its potential to become an accomplished and rare piece of taxidermy. The list containing the measurements, found on page 2 of Leyer's letter of 8 July, 1912, also evidences Sjölander's work. It was not stored in the correspondence archive, but in cover 1702 that contains material from Sjölander's work. The paper is stained and torn, and obviously served as a manual when he mounted the calf. The newspaper *Göteborgs-Tidningen* interviewed Sjölander while he was working on the mount, and reported that he now wished to recreate a small elephant herd in the museum, a cow, a bull, and the calf.[13] It is therefore possible to establish a connection between Magnus Leyer, who fired a shot that should not have been fired into a Northern Rhodesian elephant cow with a calf in 1910, and David Sjölander, who almost four decades later killed an old elephant bull on the opposite side of the African continent.

Negotiations and Preparations

Sjölander had intended to go to Africa in 1934, but was forced to cancel the journey because the museum was short of money.[14] A subsequent attempt was prevented by the Second World War. "Sjölander's favorite dream is at some point to adorn the Mammal Hall with a full grown male elephant mounted by himself, but knowing how badly such hides are handled by amateurs, he will shoot the animal himself and supervise its preparation". These are the words of Orvar Nybelin, written for the occasion of Sjölander's 60th birthday in 1946.[15] Two years later, Sjölander, arguing for collecting the elephant personally, describes taxidermy as tangential to art:

It's not like buying a fur coat for your wife when you procure an animal hide for a museum. The person who is to work on the hide should be present from the moment the animal has been shot, participate in the handling of it from the beginning, preserve it, and study the structure of the animal. The work of the taxidermist resembles as well that of the sculptor.[16]

Sjölander fought for his elephant by emphasizing the exclusive expertise of the taxidermist, the importance of knowledge through seeing and doing, the ability to effectively freeze movement and form based on field experience, and perhaps most importantly, the power to distinguish and reproduce the idiosyncratic features of an animal.

On 19 December, 1947, the museum board agreed that Sjölander should be allowed to travel to South-West Africa in 1948 and 1949 in order to collect for the museum, for a maximum period of ten months. He would retain his wage during the expedition, but the journey would be made without any costs or responsibility to the museum. His vacation should be included within the ten month period.[17] The museum already had a connection to Portuguese West Africa in Clarence Lyon, a Swedish consul in Lobito and farm owner in Lubango. Lyon had been one of the most generous sponsors of the gorilla purchase in 1906. The expedition had come about partly as a response to an invitation from Lyon to Sjölander. Sjölander had contacted Lyon about equipment, and expenses, and the game populations in Lyon's district. Lyon reported that the elephants were big and numerous, and that all kinds of game could be collected, and recommended that Sjölander arrive in June, when the rainy season had ended and the best hunting fields could be reached without difficulty. Sjölander should bring complete camping equipment and all essential gear for collecting and preserving specimens. Lyon would meet Sjölander in Loanda, attend to the necessary details, and would have plenty of time to accompany him in the bush.[18]

In the months that followed, Sjölander was busy collecting equipment and his superiors were engaged in collecting money for the expedition. The firm B.A. Hjorth & Co., producer of Primus Bacho, donated a Primus stove and two Primus lamps. Nitroglycerin Aktiebolaget, founded by Alfred Nobel, supported the expedition with ammunition.[19] The chemical laboratory Klärre & Co saw the expedition as a promotional opportunity, and donated an assortment of their products: two litres of "Djungelolja," (Jungle Oil), five litres of Maletta DDT 3, and two kg of myrrh. In a letter to Sjölander, they noted:

The products are harmless to humans and domestic animals, which means that you can use them also in the context of foodstuffs. You can thus spray wrapped foodstuffs, sacks or bags without any harm to the contents.

In return, they asked for a photograph, "a nice picture that suits our products and the motif should be jungle vegetation in the background and yourself at work. By doing so, you will do us a great favour".[20] In a letter of thanks Sjölander confirms that he will travel with cameras, and that he is looking forward to returning their generosity by means of words and pictures.[21]

Sjölander would work with Kodak film, and was obliged to report to Victor Hasselblad, his former employer, inventor of the world famous camera of the same name, and founder of Victor Hasselblad AB, telling him how the film functioned in the African climate. Hasselblad contributed 2,500 Swedish crowns to the expedition, making him by far the most generous of the private sponsors.[22] Sjölander prepared for extensive photographic documentation of his collection practices. However, the only picture of Sjölander in Angola, kept in the museum archive, showing him in a dry and dusty landscape, does not correspond at all to Klärre's vision of the African jungle.

Three days before Sjölander's departure, he was equipped by the museum with an introductory letter composed in English, French, and German, asking "the authorities of Angola, if Sjölander could obtain a special permit to kill the species of animals desirable for his task, etc.; and that he might be allowed, unimpeded, if possible custom-free, to import the scientific equipment Mr Sjölander brings along with him – which (sic) equipment is the property of the Museum – and that Sjölander, after having done his task, might be allowed to export the equipment mentioned together with the scientific materials collected".[23] The equipment is listed on five sheets, documenting the variety of items needed for hunting, collecting, sleeping, and cooking in the field.

Killed, Flayed, and Eaten

On 16 July, 1948, Sjölander left Gothenburg with the transatlantic steamer *Gullmaren* bound for Africa. At nearly 62 years of age, time was running short for him to realise his dream of collecting and mounting an African bull elephant. He travelled alone and brought with him 20 packing cases, including weapons, ammunition, arsenic, formalin, borax, carbonate, zinc, cyanide, tent, toilet paper, DDT, a mosquito net, a variety of knives, brushes, hundreds of glass jars of different sizes, 1,000 labels, matches, potato flour, six potato peelers, and a white overall coat (his usual work attire while at the museum), to mention only some of the supplies.[24] Sjölander returned to Gothenburg on 18 May, 1949, with three and a half tons of material; as much as one third comprised hides, skeletons, and tusks.

Sjölander disembarked from *Gullmaren* in Luanda, where he tried in vain to get the necessary licenses for hunting. He then continued with a Portuguese steamer south to Lobito, where the authorities forced him to leave his equipment. Sjölander himself continued with the ship to the small town of Moçamedes, today Namibe, the starting point of his collecting. He had travelled for two months and ten days before reaching the town, and due to the local authorities' reluctance to letting him have his weapons and ammunition, another month passed before he could continue towards the promontory of the Serra da Chella, a mountain range in western Angola. To retrieve his gear he had to return to Lobito by air, an uncalculated expense.[25] He had been unable to arrive in June as recommended by Lyon, and hence had to carry out his expedition in the dry season. Lyon himself was a convalescent, and could not accompany Sjölander in the field. In late November Sjölander wrote his first letter to Nybelin from "Cacoropopo, in the bushes of Mossamedes of Huila province (sic) 28/11 1948":

Because the supplies are finished I must do an extra trip to Moçamedes to buy more necessities. I will mention though, that in this part of Africa, elephants are not to be picked from trees. There are by the way quite a lot of them, but they range far and wide. This time we failed because the gun bought at Nyborg has a bad gravitation.[26]

He continues to tell how they had come within range of one bull, but the animal disappeared after receiving one bullet between the eyes, one in the ear, and two in its side. There had been no rain in the promontory the last two years, and "the days are as hot as in the so-called hell, and the nights so cold that the Swedish State's sleeping bag is insufficient to keep out the cold. … Apart from this everything is fine, that is I was rather ill before I learned to handle the water carefully – in certain areas it is not drinkable, contains too many salts of different kinds".[27] Nybelin's dry answer expresses no concern neither for Sjölander's health nor for the wounded elephant: "I'm sorry that you lost the handsome elephant. But I hope that you by now will have succeeded in getting another one".[28] Indeed, Sjölander had succeeded.

Twenty days later, Nybelin received the news in a letter from Sjölander, written in Moçamedes:

It is with great pleasure that I announce that the museum's elephant bull lies "salted" about 14 (Swedish) miles (140 km) southwest of Moçamedes Desert, and is awaiting transport home. Even the skeleton has been taken care of. The tusks, totalling 47 kg are kept in a safe place in Moçamedes. … Because of lack of rain in this area the game is continuously migrating and hence difficult to discover. I have lost two bigger elephants (wounded). The one I have managed to keep is a "medelsvensson"[29] compared to local measures. Tip to tip (the curve) 8.77 m.

Shoulder height ca 3,30–3,40 (hard to measure exactly). The skeleton will tell. I think 3,35 is correct.[30]

The elephant was shot on 4 December, 1948, and was first displayed to the public on 28 March, 1952. Between the two dates were years of endurance and hard work. First, the elephant that measured 6 metres around the belly – "a meat mountain" according to Sjölander – had to be turned around, a task that required 50 people, a jeep, an iron chain and three hours under a blazing sun.[31] The temperature was over 50 degrees Celsius. Second, the animal was flayed, an operation that lasted until 11 p.m. Finally, the skin was treated with four kg of phenol and 100 kg of salt – "the devil's work".[32] Within one day, the meat was "devoured" by the locals hired for the expedition, whom Sjölander described as "meat mad negroes".[33] Sjölander worked for two weeks until the hide, the cleaned skeleton, and the tusks could be transported to his depot. During the preparation of the hide his right hand was severely burnt by the corrosive phenol. The big, thin ears, quite torn from a life lived in the thorny bush, dried hard as bone before the elephant was fully flayed. Rawhide can also pose serious problems for preservation, since it shrinks while drying. And it can strike back: In an interview from February 1951, Sjölander claims that he never would wrap himself in a raw elephant skin because he could then be "hugged" to death. Apparently, shortly before the interview, two Swedish elk hunters had been killed because they had used a freshly flayed elk hide as a kind of blanket.[34]

A *Taxidermist's Trophy*

There is just one photograph in the museum's archive showing Sjölander and the dead elephant in Africa. The motif seems at first glance consistent with the genre "white hunter with his trophy", a genre well represented in the museum archive. A series of elephant trophy shots, taken around 1910 by Magnus Leyer, serves as an illustrative example of the genre and its conventions. Following the practice of so many other white big game hunters of his time, Leyer used both his rifle and his camera. The trophy photographs mediated the hunter's domination of animal and nature, and served as supplements or representations of displays of the material animal remains: "Displays of the remains of 'wildlife' and 'big game' in particular were used in a range of settings – from scientific institutions to entertainment venues – to convey a variety of meanings, including the colonial prospects of the territory traversed and the manliness of the intrepid hunter" (Ryan 2000: 204).

It is important to note that when Sjölander shot his elephant in 1948, the practice of photographing dead big game as useful documentation for the taxidermist when mounting the skin, was well established. In the tenth edition of *The Sportsman's Handbook to Practical Collecting, Preserving, and Setting-up Trophies and Specimens*, Rowland Ward declared that without photography the taxidermist "could never have reached his present advanced stage" (Ward 1911: 22, cited in Ryan 2000: 208). Sjölander must have taken many photographs of the elephant in order to assist the taxidermy work which was to come. These have, however, either been misplaced or destroyed.

Fig. 45. Dead elephant with the hunter on top.

PHOTO: MAGNUS LEYER, CA. 1910. GNM_ 5220_309.

The Leyer trophy photographs of killed elephants can be classified according to three conventions. One is the newly killed animal prior to being dismembered, and in these cases often only the animal is contained within the frame. The second convention is the proud, male, white hunter together with his dead prey, smiling towards the camera with his gun leaning towards the animal. In many such pictures the hunter demonstrates the utmost humiliation of the elephant by sitting or even standing on its body. The third convention is the dismembering of the elephant, showing the process of cutting up the body and

CHAPTER 4

Fig 46. David Sjölander and the dead elephant.

displays of tusks, head and feet; messy work executed by local people. As observed by Linda Kalof and Amy Fitzgerald in their analysis of hunting magazine trophy photographs: "The elephant is a particularly popular dismembered animal" (Kalof and Fitzgerald 2003: 114).

The picture must have been taken immediately after the animal had been shot, and shows Sjölander pointing at the elephant's head: was he trying to show where the lethal bullet entered the animal? Sjölander has placed his rifle on the trunk so that it leans against the animal's head, close to the still open left eye. A closer look, however, reveals that the composition differs in many respects from the typical trophy photograph. The hunter does not pose, triumphantly smiling towards the photographer. He stands turned towards the enormous dead body, his right hand pointing at a spot on the elephant's head. Since he killed it with a bullet to the heart, he cannot be pointing at a bullet hole. Neither is there much resemblance between the stereotype of a gentleman hunter and Sjölander. With the exception of the pith helmet, he is dressed as if he had been walking in the Swedish countryside: thick, clumsy trousers held up by braces, a shirt with a white cloth in one pocket and a square object in the other, perhaps an exposure meter, and heavy shoes. He has a camera hanging by a strap over his left arm. What is he pointing at?

In order understand the photograph we must look carefully at the mounted specimen. The area of the elephant's head, at which Sjölander is pointing, exhibits a spot that is shinier than the rest of the body. I had noticed this several times while contemplating the mounted elephant, but I supposed it was due to decomposition of the skin. I found the explanation, however, in Cynthia Moss's book *Elephant Memories* that just happened to be sitting on the shelf beside my desk in the museum. Moss writes:

African elephants … secrete from the temporal glands frequently and the females more than the males. … Calves of both sexes from about six months old also secrete, but as males get older they do so less often, and as full adults they usually secrete only when they are in musth, and then the liquid seems thicker and of a different consistency. (Moss 2000 [1988]: 110)

Fig. 47. The opening of the left temporal gland.

PHOTO: ANDERS LARSSON, GNM.

Suddenly I understood the photo. Sjölander is pointing at a physio-logical detail that would be very important to reproduce, if the mount were to be truly lifelike: the bull elephant's left temporal gland. Sjölander is pointing at the gland's opening, and the shiny secretion. On the mounted skin, this phenomenon is most visible on the right temple. This is a picture of the taxidermist's trophy, and the taxidermist is already involved in the process of transforming the dead animal into a specimen.

Bulky Luggage

The elephant's bones, tusks, and hide had to be moved from West Africa to Gothenburg, together with the numerous other collected items, and the expedition equipment belonging to the museum. Were there, by the way, still 81 pegs left for the tents? What is later referred to as "Sjölander's Angola Collection", consisted of skins from 35 mammals, three complete skeletons each of an elephant, a rhinoceros, and a baboon, 59 craniums, and 61 bird skins from 33 species. There were also wet specimens: three species of snakes kept in tubes or jars; 19 tubes and jars containing lizards; two jars and a tube containing frogs; two pickled tortoises; five jars holding five species of fish; one jar of freshwater crabs, and one jar of various scorpions. A letter to Professor and zoologist Bertil Hanström at Lund University from the Gothenburg Natural History Museum clarifies that Sjölander also had assembled hypophysis material from seven mammal species, kept in glass tubes.[35] On his way to Portuguese West Africa, Sjölander had collected moths, but these were sent back to Sweden on the *Gullmaren* before he travelled on to the Huila district:

During *Gullmaren's* long waiting period on the Cameroon River outside Duala when the rainy season was at its worst, I passed the time killing moths that were seeking shelter from the rain with smoke. [Sjölander smoked a pipe]. As I said, I smoked them to death because it was impossible to get to my jars of cyanide in the cargo hold. These moths are rather reduced because of rain and being killed by tobacco. This collection is now on its way to Sweden, as is a funny creature captured on the beach at Pointe Noir, French Congo. It lies in spirits and smoke perfume, and this may explain the wonderful smell. ... Send some postcards showing the exterior of the museum. Surely it will impress the Portuguese here in Portuguese West Africa.[36]

The luggage was unstable, fragile, and bulky. Sjölander writes to Nybelin that the bones of the elephant were so huge, that some had to be sawed.[37] The skeleton kept in the museum's storeroom for bones proves that this had been done. Moving jars, tubes, bones and hides safely from the inland to the coast must have put Sjölander to the test, physically as well as mentally. As far as I can discern from the final letter from Africa, written on 30 December, 1948, Sjölander planned to hire a heavy truck to transport his gear from Caporopopo to Moçamedes, and from there he intended to travel back north with the same steamer that had taken him south to Portuguese West Africa. If he ever reported to Nybelin about this last stage of the expedition, there is no evidence to be found in the museum. However, the elephant turns up for the last time on the first of June 1953, when the board considered whether the museum should reimburse three bills paid by Sjölander. One is for 300 kg of ceramic clay, the second for the transport of the clay, and the third is dated Gothenburg 10 June, 1949, when Sjölander paid 286 Swedish crowns for the cleaning and repair of a Paillard H-16 film camera. On the other side of the receipt Sjölander has written in pencil: "Camera damaged in River Caocuvular when the ferry capsized in River Caocuvular 1949".

Sjölander, the elephant's bones and hide, and the other animal remains reached Gothenburg in the middle of May 1949, but in an awkward way. Since the steamer *Gullmaren* would not port at Gothenburg, the luggage had to be discharged and declared in the Scanian town of Ystad. The hides, bones and jars were loaded onto a truck and transported the remaining 400 km up north to the museum.

What else is there to be found about the trying journey from Portuguese West Africa to Sweden? In the sources there is next to nothing. However, in an interview from May 1949, Sjölander states enigmatically: "I wonder, by the way, what the customs officials will say about the tusks I left at the custom house in Ystad".[38] To get a better understanding of what he was hinting at, we have to shift locations from the wilderness of Portuguese West Africa to the red brick castle-like building of the Gothenburg Natural History Museum.

Material Rebirth: Tusks

The African elephant's tusks are both an adornment and a marker of their sad fate. When I initiated my fieldwork in the museum, the elephant's tusks became objects of discussion and investigation: Was the elephant mounted with its own tusks, or with copies made of wood? How come that a pair of tusks is stored with the skeleton in "Benkällaren", the storage room for bones in the basement? If these were those of the elephant, and the mounted tusks wooden copies, why were the copies smaller than those in the storage room? On his list of collected specimens Sjölander has put "elephant male skin", "elephant male skeleton, male" (sic) and "elephant male tusks (only)" – does this

Fig. 48. The elephant being shaped. Note the tusks fastened to the modelled skull and another skull with tusks lying on a bench in the background. Note also the size of the manikin skeleton compared to the doors in the background.

PHOTO: VIKING FONTAINE, 1949. GNM_4576_4.

mean that he brought back a complete skeleton with tusks plus another pair of tusks?[39]

Several people touched the tusks to judge their authenticity, since ivory should feel cooler to the skin than wood, but none of the staff was an experienced "ivory toucher". Then Monica Silfverstolpe, who had been working as a taxidermist in the museum since 1965, told me that she was sure the tusks were genuine, but that they were not those of the mounted elephant. I also learned that it is extremely demanding, and almost impossible, to make a good wooden copy of tusks. After Sjölander's letters from Africa had been discovered, we learned that the tusks weighed in total 47 kg. So if the stored tusks weighed the same, the problem in identifying them would be solved. But to my disappointment the tusks in the cellar only weighed 37.2 kg. Had the pulp in the tusks weighed as much as 10 kg?

On the other hand, photos taken while Sjölander was performing the taxidermy, show the elephant's skull with tusks that look like those stored in the bone cellar, not those we see on the mounted elephant. He also spoke about tusks to the press when he returned from Africa: he had wished to shoot an elephant with bigger tusks, but since this one was an old specimen, its tusks were worn by age and from scraping the soil that was very hard after many years of drought.[40] Sjölander was licensed to shoot two elephants, and it would appear that he killed a second one. In an interview with the Swedish Broadcasting Corporation on 24 April, 1951, he said that he shot an elephant the last month he was in Portuguese West Africa. From this animal he kept the tusks, nothing else. He had not yet decided whether to use them for the mount or not, he explained, but it would be very difficult to make good copies. This second pair of tusks must have been the ones left in the customs house in Ystad, and later used in the making of the African elephant for the display in the Mammal Hall.

Material Rebirth: Body(ies)

The mounted elephant was presented to the public for the first time on 28 March, 1952, and was viewed by 5,847 visitors. Never before, or since, has a single object attracted so many people to the museum in one day (Myhrén 1983: 108). Sjölander had used Akeley's dermoplastic method: first he modelled the body in ceramic clay. Three hundred kg of clay had been ordered from a farmer, Knut Hallstén in Kvidinge, Scania, for the amount of 900 Swedish crowns. Then the clay model was covered with plaster. When the plaster dried it was taken off in pieces as negative forms to mould the manikin. The pieces of the manikin were joined together with plaster and glue, and reinforced with jute weave. Then the hide was arranged on the manikin. The biography of the dead elephant is about tusks and bones, but even more so about hide. To make the hide lighter and easier to arrange, it was shaved from 5 cm to 5 mm. To do this, the hide was soaked in a saline solution, more concentrated for each bath. Finally, it was treated with a tanning agent before being placed over the manikin. Every wrinkle was carefully folded, then reinforced by a layer of plaster, and lacquered. The seams were then camouflaged. Glass eyes were inserted. Lastly the body was painted grey and the mount was complete.

In November 1951, the Gothenburg daily newspaper *Ny Tid* reported on the elephant's authenticity: "Well, only the hide is his own, and even this has been radically shaved in the taxidermy process".[41] The day after the elephant had been shown to the public for the first time, the influential newspaper *Göteborgsposten* reported that the tusks were real, but not those of the elephant, and the local newspaper

Arbetartidningen could inform its readers that the hair on the elephant's tail had been borrowed from a Congolese relative, since this elephant was shot in the dry season and therefore lacked this important detail.[42] This effectively adds a third elephant to the mount. From where specifically were these tail hairs sourced? Perhaps from a trophy, since elephants' tails have historically been valued as such.

Big animals need space. The elephant was mounted in the Lecture Hall, a spacious room with double doors located on the same floor as the Mammal Hall. Because the completed mounted elephant would be too big to pass through the doors, the manikin was made in 52 sections (Setterberg 1989: 27).[43] After two years of work, the manikin was assembled and the mount finished in the Mammal Hall. Here the other animals and glass cases had to be rearranged to make space for the elephant in the centre, while still offering school children the opportunity to study Swedish mammals.[44] As part of this rearrangement, the skeleton of an Indian elephant gave way to the mounted African elephant.

The mounting of the hide was not only time consuming, it was physically strenuous as well. In death, the elephant struck back: one of Sjölander's assistants fell from the elephant's back while the skin was being stretched, injuring himself severely; another broke his leg during the work.[45] Sjölander himself was not present when, in 1952, he was honoured by the city of Gothenburg for a "manifold and extraordinary lifework" and particularly for his achievement in developing the museum displays.[46] Two and a half months after the unveiling of the elephant, Sjölander suffered a cerebral haemorrhage while conducting

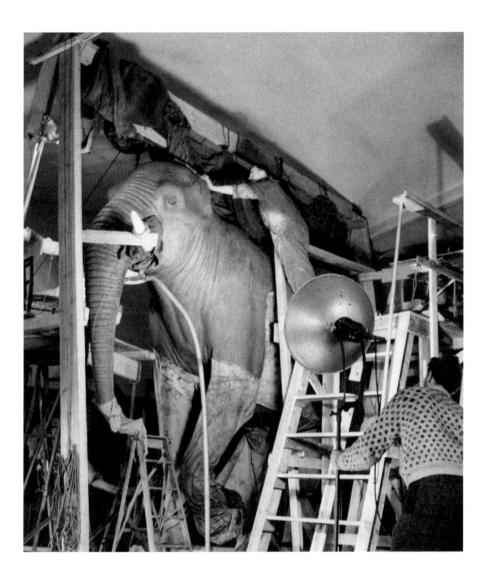

Fig. 49. The making of the clay manikin.

GNM_4576_9.

fieldwork in Swedish Lapland. Nybelin stated to *Göteborgstidningen* that the work with the elephant had been particularly strenuous to Sjölander: "Lately one has noticed that Mr. Sjölander has been very stressed. … His latest work has taken it out on him badly and his strength has been reduced after he did his expedition 1948–1949 in West Africa, where he suffered from a tropical disease".[47]

Bones: Embodied History

Few natural history museums feature mounted African elephants – many display skeletons of Indian elephants. There is a striking contrast between the physical handling of the elephant's hide and the bones, and the mediation of the mounted body and the bones. The newspaper *Göteborgsposten* interviewed Sjölander on 10 December, 1951, three months before he finished his major work. Here the story of how the elephant was successfully felled by one bullet is repeated once more, but for the first time details about its skeleton are mentioned:

One understands conservator Sjölander's joy at having succeeded in killing the huge animal with a single shot … And more so because the elephant's skeleton so clearly evidences that the animal had been exposed to several similar attacks before – among others injuries a shot had damaged an ear, and a 16mm bullet was found in its right shoulder. All in all, the skeleton was badly ravaged by bullets. [48]

Taxidermy has received considerably more scholarly attention from historians than have skeletons. Taxidermy is manipulation and ambiguity; bones are bare and bold facts. Today the elephant's skeleton is stored in "Benkällaren", hidden from the curious eyes of the public. I do not know when it was stored on these shelves or how often it has been given attention. Perhaps the last time was in the 1990s, when the bones finally were given a number in the Collectio anatomica, the register of bones.

The skeleton embodies bits of the bull's life story - its life with its own kind, and its encounters with man. Coalesced ribs demonstrate that the elephant had been in several fights, probably battling other bulls. Bullet holes and scratches show that the elephant had already been a target for hunters before its final confrontation with Sjölander. A previous bullet hole in its left shoulder blade is situated a few centimetres from the edge of the shoulder, meaning the elephant had barely escaped with its life when it sustained this injury; a few centimetres across would have meant a fatal shot to the heart. Sjölander's shot, however, ran free of the shoulder blade and struck the heart directly. Ironically, the skeleton echoes the drama in the highland of Serra da Chella, when Sjölander had failed to kill the first bull elephant that disappeared with four bullets in its body.

The Contents of the Elephant

Few can ignore a mounted African elephant on display. But how do we see it? What kind of narrative does the museum offer the visitor who contemplates the displays? The authorised story of the elephant, and of how it was turned into a museum specimen, is based on Sjölander's interviews with local newspapers, and thus represents his version of the expedition: the elephant was killed cleanly with one shot; the heat; his toil; the meat-mad "natives" or "negroes" – the cumulative effect of these elements offer an example of how a plethora of meaning disappears when an animal is inserted into a natural history collection. As Sam Alberti observes: "Zoos and museums are engines of difference, classifying and presenting the entangled mess of the natural world in a comprehensible way" (Alberti 2011: 7).

This classification and presentation of nature neatly sets aside the natural history museum's sociocultural embedment, and imbues the museum with an air of timelessness and objectivity. Yet, following Stephen Asma's paradoxical statement that "(o)nly in death do most animals pause long enough for our analytical minds to torture some truths out of them" (Asma 2003 [2001]: 27), the elephant invites us to ask what is hidden in this pachyderm, beyond its unquestionable and obvious status as a superb piece of taxidermy. In all likelihood, the elephant is not literally a container for Sjölander's African photographs and notes. Nevertheless, it is charged with significations of the collecting practices of the late nineteenth and early twentieth century natural history museum.

The elephant was collected in order to occupy a prominent, central place in the Mammal Hall. Its main purpose was educational, to serve as a three-dimensional natural history illustration, so exquisitely executed that visitors would see it as impressively lifelike. I have not been able to find any explicitly scientific reasons for Sjölander's

Fig. 50. The African elephant in the Gothenburg Natural History Museum.

PHOTO: ANDERS LARSSON, GNM.

223

expedition. Rather, the elephant represents a collecting practice aimed at "filling the gaps" or "completing the collections". The idea of a mounted elephant had been manifest in the museum since in 1904, when Leonard Jägerskiöld included an enlarged photograph of an elephant in the exhibition. Sjölander's long-held dream of collecting and mounting an African elephant was the last and most important element of his taxidermy work, and legitimized through museum policy in order to attract a public fascinated by new and rare animals.

When the board of the Gothenburg Natural History Museum met on 4 December, 1947, to discuss item 12 on the agenda, "Taxidermist Sjölander's expedition to South West-Africa", the rationale for the expedition was that "… it would be of great value to supplement certain collections with species from the West African fauna, for instance an elephant and a rhinoceros, which are now lacking in the public division".[49] Sjölander himself stated to the press before he left Gothenburg: "I will take whatever I get down there. … I will not be dedicated to a specific field of research, but collect mammals, birds, and insects".[50] Furthermore, sending Sjölander to Africa was the cheapest way to acquire an elephant. The African expedition cost 20,000 Swedish crowns; an elephant skin would have cost 12–15,000 crowns, with an extra charge for the tusks.[51]

The efficient killing of an elephant requires knowledge of the animal's anatomy, as well as familiarity with weapons and ammunition. In his memoirs, Jägerskiöld characterises Sjölander "to be an extraordinary skilled shot and a good hunter with experience even from China and the highlands of Tibet"[52] (Jägerskiöld 1943: 485). "Those of us who knew him see with our inner eye the hunter and the outdoor person, the film photographer, the travelling researcher – and in a certain way as a synthesis of all this – the master taxidermist of this very diverse and demanding profession, certainly not surpassed by anyone on this side of the Atlantic, if anywhere at all", Orvar Nybelin wrote in his obituary of Sjölander.[53] Both of them point to the importance of hunting as one of a taxidermist's required skills in the first decades of the twentieth century – not only within a minor Swedish natural history museum but in natural history museums in general. Yet, Sjölander's skill as a hunter hardly counterbalanced the problems that accumulated during his collecting in Angola.

When comparing Sjölander's expedition to Johannes Fabian's analysis of "the solitary European leading his caravan" in Central Africa in the period 1875–1910, several emerging threads appear. In *Out of Our Minds* Fabian tells an anti-myth story about some of the pioneers of European anthropology in which contradictions, incongruities, madness and ecstasies are emphasised (2000: 272). He debunks the myth of scientific collecting as individual, controlled, heroic, and rational, a practice aimed strictly at fulfilling a goal defined by the instruction of authorities, who are ignorant of the social and cultural conditions of African localities: "What should have been a matter of carrying out well-circumscribed tasks almost always turned into a battle of mere survival, with death the outcome almost as often as not" (Fabian 2000: 276). Dependence, says Fabian, was what the European traveller mostly experienced; he was dependent on funding, on all sorts of equipment, on people – and on his own health and body (2000: 276).

In his letters to Nybelin, Sjölander repeatedly complained about shortage of cash; in 1948 it took six months to transfer money from Sweden to Portuguese West Africa. In a letter dated 12 December, 1948, Sjölander expressed his disappointment and irritation at being permanently short of money, a situation which added stress to already strenuous days in the wilderness:

That cash deficit should occur now after I alerted my superiors more than two months ago! I had certainly hoped that those in charge would have shown interest in my case so that this detail would have been arranged. At any rate the gentlemen have had enough time to fix the case. I left Gothenburg on 16 July nearly half a year ago! The gentlemen have been sleeping all the time! I have not had a tranquil moment after I arrived here because of the uncertainty about the money, worrying if it would arrive in time or not.[54]

There were other problems piling up, too. Sjölander was dependent on the local authorities in order to be licensed to travel with a gun and ammunition, on farmer and hunter Vasco Ferreira to guide him, and on the locals to help with his voluminous luggage. His supposed helper, Clarence Lyon, could assist him neither in his meetings with the authorities, nor in the hunting fields. When Sjölander had at last finished with the local bureaucracy, field collecting was frequently interrupted by more mundane work. There was constantly a need for food, "… and the negroes are voracious meat eaters, so it was troublesome to constantly have to shoot for food".[55] Sjölander is here referring to the locals who did the heavy and necessary labour during the expedition. Donna Haraway states that "(t)he great halls of the American Museum of Natural History would not exist without the labor of the Africans" (Haraway 1989: 52); similarly, the splendid African elephant in Gothenburg would not exist without the toil of the Angolese locals.

Second World War turned Sjölander's African expedition into something of an anachronism.[56] This is perhaps most evident when his undertaking is compared with the Swedish deep-sea expedition aboard the *Albatross* (1947–1948), initiated by the Swedish Natural History Museum in Stockholm and including ichthyologist Orvar Nybelin as one of its members. The *Albatross* expedition was the second largest Swedish research expedition after Nordenskiöld's *Vega* expedition (1878–1880) through the North East Passage, and is considered to be one of the important undertakings in the history of oceanography. The *Albatross* circumnavigated the earth, and inspired newspaper headlines when she called at foreign ports. The crew comprised men of specialised scientific expertise; when they entered the quay Packhuskajen in Gothenburg on a sunny day early in October 1948, the crowd applauded, a brass band played, and the 'landshövding' was present.[57]

The expedition was represented the following year in the exhibition "With *Albatross* across the Ocean" at the Maritime Museum in Gothenburg. Here the public could learn that the brilliant results stemmed from a national initiative, which combined Swedish science, Swedish engineering, and Swedish seaworthiness.[58] The week before the opening of the *Albatross* exhibition, Sjölander had arrived in

Gothenburg with his truck full of specimens. No crowds, no brass band, and no "landshövding" welcomed him home.

Finally, the elephant signifies both its own species and the generic mounted animal. The elephant was Sjölander's final work and his *magnum opus*, and it ruined his health – he died in November 1954. Elephants are extremely difficult to mount, not only because they are huge and their hides heavy to handle and arrange, but also because their skin is loose and hairless, meaning the seams cannot be easily hidden. Sjölander's elephant mount, his dream materialised, will remain a testimony to his skill as a taxidermist as long as mounted animals attract spectators.

Among those clippings in the museum's scrapbook concerning the elephant mount's 1952 debut, there is only one critical voice to be found. In his article "Beastly wax cabinets" published in the Stockholm newspaper *Expressen*, Harald Hammer is presented as a spokesman for nature displays to come. Hammer asks if mounted animals "have any legitimacy in today's museums, are they strictly speaking amusement park attractions without any reasonable meaning? They are only troublesome and take up room from other and more precious items".[59] Apart from mounted specimens of extinct species, Hammer supposed that science had no use for enormous numbers of stuffed animals, stiffly posed like people in old photograph albums. Animals in nature move, in contrast to the artificial stillness of mounted animals. It was time to get away from the mausoleums of natural history, and initiate a radical rejection of old-fashioned museums. Hammer advocated films that would teach children to recognise animals and birds from their movements and sounds. It must have been destiny's irony and life's injustice that Hammer's criticism was provoked by a mount made by the award-winning nature photographer Sjölander, a pioneer of Swedish nature film. However, time has shown that Hammer's sentiments were misguided: the stuffed animals of natural history museums retain pedagogical value.

Of what value is Sjölander's elephant, apart from being an outstanding piece of taxidermy? The elephant inevitably links the Gothenburg Natural History Museum with the faunae of Portuguese West Africa. Today, the African elephant is an endangered species. The population has decreased dramatically since 1945, although several national parks and wildlife reserves have been established. The elephants in Angola are threatened with extinction as a consequence of the civil war and poaching. Elephants have been donated and transported to Angola from South Africa and Botswana in order to re-establish an elephant population.[60] If, or when, the *Loxodonta africana* go extinct, the specimen in Gothenburg will be of greater value, comparable to the mammoth. But when we pose the question in the plural, what are the values of Sjölander's elephants – two killed, two wounded, and a fifth of whom we know nothing other than that it "gave" its tail hairs to make the mount perfect – the Gothenburg Natural History Museum becomes one among the many agents involved in a process that has endangered the species, a process in which museum collectors have mingled with big game hunters and poachers. Is this too harsh a judgement of a now abandoned praxis? Should we instead be grateful that Sjölander has left the Angola elephant as both a splendid representation of the species and a manifestation of his taxidermic skill? The biographies of mounted animals may be unpleasant to present to a modern public; natural history museums have fashioned a problematic vision of nature deeply embedded in history, society and culture. But would we prefer to be without the elephant, the gorilla, Monjet, and the walrus? Good taxidermy also represents a fascination with the animals, their beauty, their variation, and their otherness – a will to postpone nature's impermanence by means of the taxidermist's expertise.

Notes

1 Portuguese West Africa is also called Angola in the sources.

2 "Europas största elefant är klar för sminkning noshörning snart i tur", *Ny tid*, 09.11.1951; "Museielefanten 'invigd'", *Göteborgs Handels- och Sjöfarts-Tidning*, 29.03.1952.

3 Elephants in Europe: A brilliant example of a highly featured elephant was Hanno, a young Indian elephant presented to Pope Leo X in 1514 as a gift from the Portuguese king Manuel I, see Bedini (1997). For the history of the elephant in Western culture, see for instance Altick (1978) about Chunee; Ritvo (1987) about Jumbo and about British big game hunting in Africa; Turchetto (2004) and Thorsen (2009) about the Indian elephant of Venice; Alberti (2011) about Maharajah in the Manchester Museum and Sutcliffe, Rutherford, and Robinson (2011) about Sir Roger the Elephant in Kelvingrove Museum, Glasgow; Rothfels (2002) about collecting African elephants, (2008) about changing perceptions of elephants and (2013) about elephant taxidermy and entropy; Roberts (2002) about conceptions of the elephant during the French Enlightenment.

4 "Denna montering blev krönet på hans livsverk som konservator och samtidigt vår skådesamlings största prydnad"; "... oöverträffad mästare". *Göteborgs naturhistoriska museum. Årstryck 1953*, 6.

5 Letter from L.A. Jägerskiöld to the Board of the Gothenburg Museum 29.12.1927.

6 "... med sina femtiofyra fällda elefanter är den störste nu levande elefantjägaren"; "att han skjutit åtskilligt storvilt i sin dag och också bekräftar elefantsiffran, men ... är mycket noga med att påpeka att han aldrig varit yrkesjägare utan egentligen mera av en händelse kommit att nedlägga en hel del vilt under de vidsträckta resor han som brittisk ämbetsman företagit genom sitt stora fögderi". "Elefantsnabel smakar bra, tämja zebror besvärligt", *Handelstidningen*, 30.11.1934.

7 Letter from M. Leyer to L.A. Jägerskiöld 17.01.1912. Leyer spelled his family name with a "y" in his letters written in English to Jägerskiöld; in the museum photo archive and in newspaper articles the name is often spelled Leijer.

8 Letter from M. Leyer to L.A. Jägerskiöld 11.03.1912.

9 Letter from M. Leyer to L.A. Jägerskiöld 18.01.1911.

10 Letter from M. Leyer to L.A. Jägerskiöld 18.01.1911.

11 Letter from M. Leyer to L.A. Jägerskiöld 08.07.1912.

12 Letter from M. Leyer to L.A. Jägerskiöld 08.07.1912.

13 "40-årig nyfödd elefantunge föds på nytt i Naturhistoriska", *Göteborgs-Tidningen*, 03.12.1947.

14 Oral information from Torkel Hagström.

15 "David Sjölanders älsklingsdröm är att en gång få pryda Naturhistoriska museets däggdjurssal med en av honom monterad fullvoxen elefanthane, med med kännedom om hur illa alla sådana hudar brukar behandlas av icke fackmän vill han själv nedlägga djuret och leda dess tillvaratagande". "David Sjölander 60 år", *Göteborgs Handels- och Sjöfarts-Tidning*, 16.11.1946.

16 "Det är inte som att köpa en päls till frun, när man ska anskaffa en djurhud för ett museum. Den som skall montera skinnet skall helst vara med från det ögonblick djuret skjutes, vara med om att ta hand om det från början, preparera det och ta en titt på djurets struktur. En konservators arbete liknar också på en skulptörs". "Skaffa elefanthud är inte så enkelt som fruns pälsköp", *Ny Tid* 29.04.1948.

17 Protokoll, hållet vid sammenträde med styrelsen för Göteborgs museum den 19. december 1947. *Gbgs museistyrelsesprotokoll 1947–1948*.

18 Letter from C. Lyon to D. Sjölander 09.10.1947.
19 Brev from Nitroglycerin Aktiebolaget to D. Sjölander 15.06.1948.
20 "Samtliga varor äro ofarlliga för människor och husdjur, varför Ni kunna använda Eder av produkterna även ifråga om livsmedelförråden. Ni kan således bespruta förpackade livsmedel, säckar feller påsar utan att innehållet därav tager skada". "… en trevlig bild, som lämpar sig för våra varor och skulle då som motiv vilja ha en spec. djungel-vegetation som bakgrund med Eder själv i arbete, med vilket Ni skulle göra oss en stor tjänst". Letter to D. Sjölander from Klärre & co 08.06.1948.
21 Letter to Klärre & co from D. Sjölander 15.06.1948.
22 Letter to V. Fontaine from V. Hasselblad 28.06.1948. The total cost of the expeditions was 22,300 Swedish crowns. The members of the museum board changed their minds and contributed 7,000 crowns, the town of Gothenburg 6,500 crowns. The rest was given by private sponsors. *Göteborgs Naturhistoriska Museum. Årstryck 1948– 1949*, 18; Protokoll, hållit vid styrelsen för Göteborgs museum den 11. mars 1948. *Gbgs museistyrelsesprotokoll 1947–1948*.
23 Introductory letter of 13.07.1948.
24 List dated 4 July, 1948, in *Gbgss musei styrelsesprotokoll 1948–1956*.
25 Letters from D. Sjölander to O. Nybelin 26.09.1948 and 25.10.1948.
26 "På grund av att provianten är slut måste jag göra en extra resa till Moçamedes för att köpa mera förnödenheter. Jag vill herved omtala att elefanter plockas ej från träd här i denna del av Afrika. Här är ganska gott om dem emellertid men de ströva vida omkring". Letter from D. Sjölander to O. Nybelin 28.11.1948.
27 "Dagarna äro heta som i s.k. helvete, och nätterna så kalla att svenska statens sovsäckar räcka ej till för att hålla kylan borta. … Annars är allt väl dvs. Jag har varit ganska dålig innan jag lärde mig att hanskas försiktigt med vattnet – det är ej passande att dricka i vissa trakten, innehållar allt för mycket salter av div. slag". Letter from D. Sjölander to O. Nybelin 28.11.1948.
28 "Skada bara att Du gick miste om den vackra elefanthanen. Jag hoppas emellertid att Du vid det här läget har lyckats med en annan". Letter from O. Nybelin to D. Sjölander 13.12.1948.
29 "Medelsvensson" normally means an ordinary person, here it means middle-sized.
30 "B.B! Har härmed nöjet meddela att museets elefanttjur ligger "insaltad" circa 14 s mil s.w. om Moçamedes desert (sic), och väntar på möjlihet (sic) för hemresa. Även skelettet är tillvaratagit. Betorna 47 kg tillsammans är i säkert förvar i Moçamedes. … På grund av brist på regn inom detta område ha viltet varit på ständig vandring och därför svårt at finna. Jag har förlorat två större elefanter (sårade) Den jag tillvarataget är "medelsvensson" efter angola-förhållanden. Tip to tip efter kurvan 8,77 meter. Boghöjd circa 3,30-3,40 (svårighet att mäta exakt). Skelettet får avgöra. Jag tror 3,35 är riktiga". Letter from D. Sjölander to O. Nybelin 25.12.1948.
31 "Salongsgevär föga hopp mot hotfullt afrikalejon", *Göteborgs-Posten*, 25.05.1949.
32 "… ett satans göra". Letter from D. Sjölander to O. Nybelin 25.12.1948.
33 "Ett köttberg! Den slukades av negrerna på en dag. Temperaturen i solen 50 gr cl.ingen skugga, köttgalna negrer". Letter from D. Sjölander to O. Nybelin 25.12.1948.
34 "Tjuvaktig jättepojk, som fick stryk på friarstråt!", *Göteborgs-Posten*, 10.02.1951.

35 Sjölander's handwritten minutes, archive number 1712; letter from Gothenburg Natural History Museum to B. Hanström 22.10.1949.

36 "Under *Gullmarens* långa väntetid på Kamerun-floden utanför Duala just under värsta regntiden därstäds fördrev jag tiden med att röka ihjäl nattflygare vilka sökte en fristad under regnet. Som sagt jag rökte ihjäl dem därför att mina cyanburkar befannt sig outkomliga i lastrummet. Fjärilarne ifråga äro åtskilligt avflugna på grund av regn och ihjälsandet med tobak. Denna samling är nu på veg till Sverige, likaledes en konstig figur fångad på badstranden vid Pointe Noir franska Kongo. Det ligger i sprit och rökparfym, det kanske förklarar den ljuvliga doften. … Sänd några vykort av museet, exteriören. Det kommer nog att imponera på portugiserna här nere i Angola". Letter from D. Sjölander to O. Nybelin 26.09.1948.

37 Letter from D. Sjölander to O. Nybelin 30.12.1948..

38 "Vad månne tullen säga förresten, om de elefantbetar jag lämnade kvar på på tullkammaren i Ystad". "Salongsgevär föga hopp mot hotfullt afrikalejon", *Göteborgs-Posten*, 25.05.1949.

39 Sjölander's handwritten minutes, archive number 1712.

40 "Salongsgevär föga hopp mot hotfullt afrikalejon", *Göteborgsposten*, 25.05.1949.

41 "D.v.s. det är bara skinnet som är hans eget, och till och med det är avtunnat betydligt för att kunna prepareras". "Europas största elefant är klar för sminkning noshörning snart i tur", *Ny Tid*, 09.11.1951.

42 "Museielefanten ståtar med skotthål i öronen", *Arbetartidningen*, 29.03.1952.

43 The story told in the museum says that Sjölander miscalculated the size of the elephant and therefore had to cut the manikin in pieces. It is not likely that the systematic and practical Sjölander, who knew the accurate measurements of the elephant's body, would make such a blunt mistake.

44 "Jätteelefant vållar möbleringsproblem", *Göteborgsposten*, 12.11.1950.

45 "Konservant slant från elefant", *Arbetarposten*, 29.03.1952.

46 "för mångsidig och märklig livsgärning"; "Festlig afton på Börsen Förtjänsttecknen jubilera", *Handelstidningen*, 05.06.1952.

47 "Man har på sista tiden märkt att konservator Sjölander varit mycket ansträngd … Hans senaste arbete har tagit hårt på honom och hans krafter har varit nedsatta ändå sedan han gjorde sin expedition 1948–1949 i Västafrika, där han drabbades av en tropiksjukdom". "Känd göteborgare låg sanslös vid fjällstup", *Göteborgsposten*, 16.06.1952.

48 "Man förstår konservator Sjölanders glädje over att ha lycktas å fälla det stora djuret med ett enda skott … Dette så mycket mer som elefantens skelett tydligt bar spår av att djuret tidigare utsatts för åtskilliga attacker av samma slag – bl.a. hade ett skott träffat i ena örat och ett annat, en 16 mm tjock kula, satt kvar i högra skuldran. Skjelettet var mycket illa åtgånget av gevärskulor även i övrigt". "Tjuvaktig jättepojk, som fick stryk på friarstråt!", *Aftonposten*, 11.01.1951.

49 "… vore det synnerligen värdefullt att få vissa samlingar kompletterade med representanter ur den Väst Afrikanska faunaen, t.ex. elefant och noshörning, vilka nu saknas i skådesamlingen". *Naturhistoriska museet Nämdens protokoll 1944–1958*.

50 "Jag skall ta, vad jag kan få där nere … Något spesielt forskningsfält kommer jag inte att ägna mig åt, utan däggdjur, fåglar ock insekter kommer jag att samla". "David Sjölander väntar nu bara på

lämplig båt", *Handelstidningen*, 29.06.1948.

51 "Naturhistoriska museets jätte-elefant invigd", *Göteborgsposten*, 29.03.1952.

52 "Till konservatorns yrke hör att även vara jägare. David Sjölander är en ovanligt skicklig skytt och god jägare med erfarenhet även från Kina och Tibets högland" (Jägerskiöld 1943: 485).

53 "Vi som kände honom se för vår inre syn jägaren och friluftsmänniskan, naturälskaren, Lapplandskännaren, filmfotografen, forsknings-resanden och - i viss mån som en syntes av allt detta - den mästerlige konservatorn, i detta lika mångsidiga som krävande yrke säkerligen ej överträffad av någon på denna sidan av Atlanten om ens någonstädes". "David Sjölander död", *Handelstidningen*, 18.11.1954.

54 "Att kassabrist skulle inträffa vid denna tidpunkt har jag meddelat merän för två månader sedan, och jag hade verkligen hoppats att vederbörande kunde visat så mycket intresse (sic) för saken att den detaljen ordnats. Herrerna har alla fall havt god tid att ordna biffen. Jag avreste 16/7 från Gbg. Snart ett halvt år! Under tiden har herrerna sovit! Jag har inte havt en lugn stund sedan jag kom ut här på grund (sic) denna osäkerhetskänsla för pengarna, om de kommer til tid eller ej". Letter from D. Sjölander to O. Nybelin 25.12 1948.

55 "… negrerna är några förfärliga köttätare, så det var rätt besvärligt att skjuta mat till jämnt". "Salongsgevär föga hopp mot hotfullt afrikalejon", *Göteborgsposten*, 25.05.1949.

56 Even as late as in 1952, Sjölander's successor, Björn Wennerberg, went to East Africa to collect specimens.

57 "Grant och högtidligt när Albatross kom hem", *Göteborgs Handels- och Sjöfarts-Tidning*, 04.10.1948

58 "Med Albatross över haven", *Göteborgs Handels- och Sjöfarts-Tidning*, 17.05.1949 and "Albatross populariserar sina expeditionsbragder", *Göteborgs Morgonpost*, 27.05.1949.

59 "Frågan är emellertid om sådan där förevisning verkligen hör hemma på våra dagars museer. Egentligen är de väl rena tivoliattraktioner som inte längre tjänar någon förnuftig mening? De förorsakar bara besvär, tar plats från andra, värdefullare förvärv". "Djuriska vaxkabinett", *Expressen*, 10.06.1952.

60 http://news.nationalgeographic.com/news/2001/09/0904_TVelephantlift_2.html. Retrieved 9 January, 2012.

Animals, Society, and History

The four biographies outline a collecting policy that the Gothenburg Natural History Museum shared with other Western natural history institutions in the first half of the twentieth century. The objects came to the museum as gifts, purchases, fieldwork, and exchanges. While showing that the collecting history of the Gothenburg Natural History Museum, its "politics of acquisition", equals that of its Western sister institutions, the trajectories of these animals also exemplify that "every museum object has the potential for a rich life history" (Alberti 2009: 110–111). The gorilla and the walrus were both bought by the museum, but while the former was bought as a stage in a planned collecting policy, the latter was bought as a result of a random coincidence. Monjet the monkey was a gift to the museum, but her biography can just as easily be read as a bizarre story about a pet that was re-homed in a natural history museum. The history of the elephant may be categorized as part of museum fieldwork, but the dramatic entanglement of the life histories of Sjölander and the elephant reduces his collecting of the other specimens to a backdrop.

Many of the mounted mammals in the Gothenburg Natural History Museum are old specimens, documenting a collecting practice which today is deemed illegal and unethical. For this reason, it can be uncomfortable to present their biographies to the public. Natural history collections are not neutral but deeply embedded in history, society, and culture. The lowland gorilla, the walrus, the Tonkean macaque, and the African elephant are today threatened species, and national and international legislation has attempted to regulate their existence.[1]

Live and dead animals became lucrative goods in the period stretching from 1880 to 1930. The establishment of zoological gardens and natural history museums in Europe and the USA created an expanding market for this special category of goods. The period has been called "the 'heyday' of natural history", a time when specimens from all over the world flowed into the capitals of the British Empire and other important harbour cities (Alberti 2009: 31). Many were sold to British museums and collections, but some travelled even further. One of these was the gorilla which ended in Gothenburg.

The collection of exotica and natural specimens followed in the wake of colonization and imperialism. Several researchers have proved the link between serving in the colonies and the interest in hunting big game. Men working for the British colonial administration played an important role in collecting specimens from new territories that were placed under British rule during the first half of the 1800s (Browne 1996: 306, Ritvo 1986: 249). Foreigners, such as Magnus Leyer, also served in British colonies, and sold specimens directly to contacts in their homelands. In the German colonies, "the animal-catching business was taken over by colonizers and professionalized in much the same way that other colonial industries had been" (Rothfels 2002: 53). The German Carl Hagenbeck organized the collection of live animals

CONCLUSION

for his zoological garden in Hamburg, also using it as a transit station for animals which were sold on to other zoological gardens around Europe (Rothfels 2002). Several of the specimens in the monkey cabinets in the Gothenburg Natural History Museum had been bought from Cuneo's menagerie, thus establishing a link between the museum and the menageries' travelling collections of live exotic animals.

Scientists and taxidermists connected to the museum world were also active collectors. Already at the end of the nineteenth century wildlife was dwindling and species were vanishing, and in *Taxidermy And Zoological Collecting – A Complete Handbook For The Amateur Taxidermist, Collector, Osteologist, Museum-Builder, Sportsman And Travellers*, published in 1899, William T. Hornaday (1854–1937) warned collectors of natural specimens: "Every large terrestrial mammal species on earth is being killed faster than it breeds!" (Hornaday 1899: vii). It would soon be too late to collect, for example, walrus, elk, and manatee. Anyone who wanted a specimen of these threatened species should set off immediately to collect them. It was already too late to collect a wild American bison (Hornaday 1899: ix). At this time Hornaday was the director of the newly established New York Zoological Park, today the Bronx Zoo. Being a naturalist, taxidermist, conservationist, and collector, Hornaday was, on the one hand, an American protagonist in moving animals from their natural habitats to museums and zoos, and, on the other hand, a pioneer in the efforts to preserve American wildlife.[2] Walter Kaudern was one of the many scientists who travelled and collected rare animals. Monjet the monkey was captured as an infant and might have ended in a zoo like thousands of other animal infants before it. By chance she was picked up by a Swedish zoologist on a collecting expedition with his family in the Sulawesi inland. Due to her peaceful disposition, she was accepted into the family as a pet, and even if her fate was captivity, she had greater freedom and more social stimulus than she would have had in a zoological garden.

The place from where the animals had been collected does not appear to have impacted their value as museum specimens. Whether they had been killed in their natural habitat, as the elephant had been, had perished, as the walrus had because it lost its way, or had been a pet like Monjet, they were all transformed into natural objects.

In Victorian Great Britain there was general agreement that specimens killed in nature, particularly large carnivores, were preferable for scientific study and display over animals that had been kept in captivity. Animals in captivity did not have the opportunity to develop bones and muscles "in a violent manner", and it was therefore claimed that they did not have fully developed skeletons (Ritvo 1987: 248). Similar issues cannot be traced in annual reports written by Jägerskiöld or in his correspondence with Rowland Ward Ltd. What was essential for Jägerskiöld was that the specimens to be displayed to the public should be first class with respect to their fur and stance, and they should preferably be male; it was not important whether or not the animals had been killed in their natural habitat. Nor has it been possible to find protests about whether or not animals that had been killed illegally, such as the okapi and the one month old elephant calf, should be added to the collections. Rather they may be considered as a form of "museum trophies", attractions that would draw the public to the museum exhibitions.

A Large Red Heart

The animal biographies demonstrate that dreams, emotions, and chance events have been formative even to something as instrumental as the creation of museum collections. The biographies are narratives about the death of animals and persons: Hilmer Skoog died while working to imbue the walrus with an afterlife, David Sjölander exhausted himself in his work on the elephant. It is impossible to understand the materiality of the preserved animals without including the taxidermist. The most important link between a flat hide and a first-class museum animal is the taxidermist and his or her skills. According to the philosopher Stephen Asma, there is, in the metaphorical sense, a smell of death permeating natural history museums (Asma 2003 [2001]: 37). Good taxidermy creates trust, and contributes to alleviating the anxiety created by the encounter with droves of dead animals.

The transformation of an individual animal into a piece of taxidermy, an object where the only thing stemming from the animal body will often be the hide, represents the ultimate objectification of the animal. Mounted animals are objects, but each individual animal is unique: "The skin, our largest organ, is the frontier between the inside and the outside of our bodies. It is one of our chief defining factors and yet it is unique; the pigmentation, the texture, the genetic characteristics, the hair, the fur, the freckles or the feathers mark out each species, each family, and indeed each individual creature. Skin is valuable; it is eaten, worn, upholstered, and fetishistic. It is the raw material of taxidermy" (Eastoe 2012: 10). Taxidermy is a profession

Fig. 51. Lynx mounted by Monica Silfverstolpe.

PHOTO: ANDERS LASSON, GNM.

where one attempts to create life in dead animals, life understood as giving the animal character, an expression so lifelike that for a second we are tricked into forgetting that we are looking at a dead object. The taxidermist maintains not only a close, but also often an empathetic relationship to nature and to living as well as dead animals.

If you enter the original entrance to the Mammal Hall, you will see in the first cabinet on the right a small lynx. This lynx is a male aged nine months, hit and killed by a car in the sparsely populated district of Härjedalen in Sweden on 26 December, 1989. The driver, who was from Gothenburg, brought the dead lynx back to the city and presented it to the museum, where Monica Silfverstolpe worked on the mount for four months. She made the entire assembly alone, falling so much in love with the lynx that she placed a large red cardboard heart inside it. Today she is retired, but still visits the museum often. Each time she goes to see the lynx. What, then, is this lynx mount? It is primarily intended to represent the only living big cat in Sweden, *Lynx lynx*, thus it is an important element in the exhibition of national fauna. As road kill, the lynx is a monument to the numerous wild animals that each year clash with civilization, represented by its cars and trains. It is a good piece of taxidermy that succeeds in convincing museum visitors that it is a lynx, not merely a stuffed hide. Finally, it is an emotional object which houses a heart, a declaration of love by the taxidermist to the animal that once was, the lynx she never saw alive, and which she has given not just a new body, but also a new heart.

From 1900 to 1950 journalists from Gothenburg newspapers regularly visited the museum taxidermy workshop. The collecting histories and the work of taxidermists to give dead animals a lifelike form became exciting narratives. They ended well – bearing in mind that the naturalized animal is returned to its form and characteristics and can be displayed to the public without bullet holes or other injuries. I have not been able to pinpoint the exact place where Sjölander's bullet penetrated the elephant skin or where the walrus was injured when it came on to the shore at Skagen. The blind eyes have been replaced with new ones made of glass. Preserved animals' mimicry of life evokes reactions and sentiments depending on who sees the animal, the historical epoch it is seen in, and where the animal is seen. But once it is moved out of the glass case, illusion and meaning will be at stake. The mounted animals that have been studied for this book are anchored materially in the museum institution and scientific practice. It has been my aim to set them free by telling their stories.

Notes

1 The first international measure to control the trade in wild animals came in 1973, when 80 nations signed CITES, the Convention of International Trade in Endangered Species of Wild Fauna and Flora.

2 In 1905 William T. Hornaday founded the American Bison Society with Theodore Roosevelt to promote the preservation of the species.

References

Akeley, C. (1923). *In Brightest Africa*. New York: Garden City.

Alberti, S.J.M.M. (2005). "Objects and the Museum." *Isis* 96 (4), 559–571.

Alberti, S.J.M.M. (2008). "Constructing Nature behind Glass." *Museum and Society* 6 (2), 73–97.

Alberti, S.J.M.M. (2009). *Nature and Culture. Objects, disciplines and the Manchester Museum.* Manchester: Manchester University Press.

Alberti, S.J.M.M. (2011). "Maharajah the Elephant's Journey". In: Alberti, S.J.M.M. (ed.). *The Afterlives of Animals*. Charlottesville and London: University of Virginia Press, 37–58.

Alberti, S.J.M.M. (ed.). (2011). *The Afterlives of Animals*. Charlottesville and London: University of Virginia Press.

Allin, M. (1998). *Zarafa – A Giraffe's True Story. From Deep in Africa to the Heart of Paris*. New York: Dell Publishing.

Altick, R.D. (1978). *The Shows of London*. Cambridge, Massachusetts and London: Belknap Press of Harvard University Press.

Amnehäll, J. (1997). "Prakt och Prydnader. Glimtar från den indonesiska ön Sulawesi". In: *Power & Gold*. Göteborg: Göteborgs etnografiska museum, 8–24.

Andersson, D.T. (2001). *Tingenes taushet, tingenes tale*. Oslo: Solum Forlag.

Anonymous (1927). "En valross skuten i Bohuslän". *Fauna och Flora. Populär Tidsskrift för Biologi* 22 (1), 43–44.

Asma, S. (2003 [2001]). *Stuffed Animals and Pickled Heads: The Culture and Evolution of Natural History Museums*. New York: Oxford University Press.

Astley-Marberly, C.T. (1975 [1960]). *Animals of East Africa*. Nairobi: Hodder and Stoughton.

Bedini, S.A. (1997). *The Pope's Elephant*. Manchester: Carcanet Press.

Bernström, J. (1956). "Apor". *Kulturhistorisk leksikon for nordisk middelalder fra vikingtid til reformasjonstid.* I, 171–173. Oslo: Gyldendal norsk forlag.

Bevan, W.L. and Phillott, H.W. (1873). *Mediæval Geography. An Essay in Illustration of the Hereford Mappa Mundi*. London: E.K. Jakeman and J. Jones.

Bodry-Sanders, P. (1998). *African Obsession. The Life and Legacy of Carl Akeley*. Jacksonville: Batax Museum Publishing.

Brassel, J. (1908). "Dr. med. Georg Albert Girtanner. Sein Lebensbild". *Separat Abdruck aus dem Jahrbuch 1907 der St. Gallischen Naturwissenschaftlichen Gesellschaft*. St. Gallen: Buchdruckerei Zollikofer & Cie.

Brenna, B. (2013). "The Frames of Specimens: Glass Cases in Bergen Museum Around 1900." In: Thorsen, L.E., Rader, K. and Dodd, A. (eds.). *Animals on Display: The Creaturely in Museums, Zoos and Natural History*. University Park, Pennsylvania: Pennsylvania State University Press, 37–58.

Browne, J. (1996). "Biogeography and Empire."

In: Jardine, N., Secord J.A. and Spary, E.C. (eds.). *Cultures of Natural History*. Cambridge: Cambridge University Press, 305–322.

Brusewitz, G. (1993). *Natur och illusion. Biologiska museet*. Stockholm: Informationsförlaget.

Carlén, O. (1869). *Göteborg. Beskrifning öfver staden och dess närmaste omgifningar. Ny handbok för resande*. Stockholm: Oscar L. Lamms Förlag.

Clark, C.A. (2008). *God – or Gorilla. Images of Evolution in the Jazz Age*. Baltimore: Johns Hopkins University Press.

Cokinos, C. (2009 [2000]). *Hope is the Thing with Feathers. A Personal Chronicle of Vanished Birds*. London: Penguin Books.

Connif, R. (2011). *The Species Seekers. Heroes, Fools, and the Mad Pursuit of Life on Earth*. New York and London: W.W. Norton & Company.

Damsholt, T., Simonsen, D.G. and Mordhorst, C. (eds.). (2009). *Materialiseringer. Nye perspektiver på materialitet og kulturanalyse*. Aarhus: Aarhus University Press.

Daston, L. (ed.) (2004). *Things That Talk. Object Lessons from Art and Science*. New York: Zone Books.

Drivenes, E.-A. (2004). Ishavsimperialisme. In: Drivenes, E.-A. and Jølle, D. (eds.). *Norsk Polarhistorie*. II, 175–259. Oslo: Gyldendal.

Du Chaillu, P.B. (1861). *Explorations And Adventures In Equatorial Africa: With Accounts Of The Manners And Customs Of The People, And Of The Chase Of The Gorilla, The Crocodile, Leopard, Elephant, Hippopotamus, And Other Animals*. New York: Harper & Brothers.

Eastoe, J. (2012). *The Art of Taxidermy*. London: Pavilion.

Eckhart, G. and Lanjouw, A. (2008). *Mountain Gorillas. Biology, Conservation and Coexistence*. Baltimore: Johns Hopkins University Press.

Everest, S. (2011). "Under the Skin: The Biography of a Manchester Mandrill." In: Alberti, S.J.M.M. (ed.). *The Afterlives of Animals*. Charlottesville and London: University of Virginia Press, 75–92.

Fabian, J. (2000). *Out of Our Minds. Reason and Madness in the Exploration of Central Africa*. Berkeley and Los Angeles: University of California Press.

Flinterud, G. (2013). *A Polyphonic Polar Bear. Animal and Celebrity in Twenty-first Century Popular Culture*. PHD Dissertation. University of Oslo: Faculty of Humanities.

Fåhræus, G. (1983). "Olof Fåhræus – ämbetsmannen, entomologen, museiskapararen". *Göteborgs Naturhistoriska Museum 150 år. Årstryck 1983*, 15–23.

Genesko, G. (2005). Natures and Cultures of Cuteness. *Invisible Culture. An Electronic Journal for Visual Culture*. http://www.rochester.edu/in_visible_culture/Issue_9/genosko.html.

Gustafson, K. (2009). *Strand och hav*. Göteborg: YC bokförlag.

Gütebier, T. (1995). "Zur Erinnerung an den Altmeister der Dermoplastik Karl Kaestner 19.1.1895 bis 10.10.1983". *Der Präparator* 41 (1), 3–16.

Gütebier, T. (2011). "Das Bamberger Quagga und dessen Neuafstellung durch Karl Kaestner im Jahre 1969". *Der Präparator* 57, 54–65.

Haraldson, S. and Källgård, A. (1997). *Öfolk, snöfolk och nomader – resor i världens obygder*. Stockholm: Carlssons.

Haraway, D. (1989): *Primate Visions. Gender, Race and Nature in the World of Modern Science*. New York and London: Routledge.

Hedqvist, E. (2009). Varats och utvecklingens kedja.

En museinaturhistorisk utställning i Göteborg 1923–1968. *Papers in Museology 6.* Umeå Universitet.

Henning, M. (2006). *Museums, Media and Cultural Theory.* Maidenhead: Open University Press.

Hornaday, W.T. (1899). *Taxidermy and Zoological Collecting – A Complete Handbook For The Amateur Taxidermist, Collector, Osteologist, Museum-Builder, Sportsman And Travellers.* New York: Charles Scribner's Sons.

Impelluso, L. (2004). *La natura e i suoi simboli. Piante, fiori e animali.* Milano: Mondadori Electa.

Jägerskiöld, L.A. (1943). *Upplevt och Uppnåt. Ur minne, brev och loggböcker.* Stockholm: Natur och Kultur.

Jensen, A.S. (1927a). "Hvalrossen ved Skagen og dens Vandringsveje." *Naturens Verden.* June–July, 1–5.

Jensen, A.S. (1927b). "On a Walrus (*Trichechus rosmarus L.*) which has visited Denmark and probably four other European Countries." *Videnskabelige Meddelelser fra Dansk Naturhistorisk Forening* 4, 189–193.

Kalof, L. and Fitzgerald, A. (2003). "Reading the trophy: exploring the display of dead animals in hunting magazines." *Visual Studies* 18 (2), 112–122.

Kaudern, W. (1921). *I Celebes obygder.* I & II.Stockholm: Albert Bonniers Förlag.

Kaudern, W. (1937). "Anthropological Notes from Celebes." *Ethnological Studies 4.* Göteborg: Elanders Boktryckeri.

King, J. E. (1983). *Seals of the world.* London: British Museum and Oxford University Press.

Lange, B. (2005). "Die Allianz von Naturwissenschaft, Kunst und Kommerz in Inszenierungen des Gorillas nach 1900." In: Zimmermann, A. (ed.): *Sichtbarkeit und Medium. Austasch, Verknüpfung und Differenz naturwissenschaftlicher und ästhetischer Bildstrategien.* Hamburg: Hamburg University Press. http://hup.rrz.uni-hamburg.de, 183–210.

Lange, B. (2006). *Echt. Unecht. Lebensecht. Menschenbilder im Umlauf.* Berlin: Kulturverlag Kadmos.

Lange, B. (2007). *Ein Riesengorilla auf Sankt Pauli.* http://www.thing-hamburg.de/index.php?id=491.

Larsen, P. (2011). "Reisebilder. Polfarernes vei inn i den visuelle kulturen." In: Østgaard Lund, H. and Berg, S.F. (eds.). (2011). *Norske polarheltbilder 1888–1928.* Oslo: Forlaget Press, 12–18.

Laursen, D. (1954)." Adolf Severin Jensen". *Særtryk af Meddelelser fra Dansk Geologisk Forening.* 12. Copenhagen, 521–523.

Lorenz, K. (1971). *Studies in Animal and Human Behaviour.* II. Cambridge, Massachusetts: Harvard University Press.

Lyngø, I.J. (2003). Vitaminer! Kultur og vitenskap i mellomkrigstidens kostholdspropaganda. *Series of dissertations submitted to the Faculty of Arts, University of Oslo.* No. 162. Oslo: Unipub.

Löfgren, Orvar (1985). "Our Friends in Nature: Class and Animal Symbolism." *Ethnos* 50, 184–213.

Madden, D. (2011). *The Authentic Animal. Inside the Odd and Obessive World of Taxidermy.* New York: St. Martin's Press.

Maestripieri, D. (2007). *Macachiavellian Intelligence. How Rhesus Macaques and Humans Have Conquered the World.* Chicago: University of Chicago Press.

Mathiasson, S. (1983). "Zoologen, mångsysslaren och djurskyddsmannen A. W. Malm – innstiftaren av "Sällskapet Småfoglarnes Vänner", föregångaren till Göteborgs Djurskyddsförening". *Göteborgs Naturhistoriska Museum 150 år. Årstryck 1983,* 23–33.

Maxwell, G. (1967). *Seals of the World*. London: Constable.

Milgrom, M. (2010). *Still Life. Adventures in Taxidermy*. Boston and New York: Houghton Mifflin Harcourt.

Morris, P. (2003). *Rowland Ward. Taxidermist to the World*. Lavenham: Lavenham Press.

Morris, P.A. (2010). *A History of Taxidermy: art, science and bad taste*. Ascot: MPM Publishing.

Moss, C. (2000 [1988]). *Elephant Memories*. Chicago: University of Chicago Press.

Myhrén, S. (1983). "Från kuriosakabinett till miljö-museum – om utställningsverksamheten vid Göteborgs naturhistoriska museum". *Göteborgs Naturhistoriska Museum 150 år. Årstryck 1983*, 107–113.

Netz, R. (2004). *Barbed Wire. An Ecology of Modernity*. Middletown: Wesleyan University Press.

Nussbaumer, M. (2000). *Barry vom Grossen St. Bernhard*. Bern: Simowa Verlag.

Orrhage, L. (1983). "L.A. Jägerskiöld som musei-organisatör och folkbildare". *Göteborgs Natur-historiska Museum 150 år. Årstryck 1983*, 33–41.

Owen, Richard (2012 [1865]). *Memoir on the Gorilla (Troglodytes Gorilla, Savage)*. Facsimile. Memphis: General Books LLC.

Paddon, H. (2011). "Biological Objects and "Mascotism". The Life and Times of Alfred the Gorilla". In: Alberti, S.J.M.M. (ed.). *The Afterlives of Animals*. Charlottesville and London: University of Virginia Press, 134–151.

Patchett, M., Foster, K. and Lorimer, H. (2011). "The Biographies of a Hollow-Eyed Harrier." In: Alberti, S.J.M.M (ed.). *The Afterlives of Animals*. Charlottesville and London: University of Virginia Press, 110–134.

Philo, C. and Wilbert, C. (eds.) (2000). *Animals Spaces, Beastly Places. New geographies of human–animal relations*. London and New York: Routledge.

Poliquin, R. (2008). "The Matter and Meaning of Museum Taxidermy." *Museum and Society* 6 (2), 123–134.

Poliquin, R. (2011). "Balto the Dog". In: Alberti, S.J.M.M. (ed.). *The Afterlives of Animals*. Charlottesville and London: University of Virginia Press, 92–110.

Poliquin, R. (2012). *The Breathless Zoo. Taxidermy and the Cultures of Longing*. University Park, Pennsylvania: Pennsylvania State University Press.

Redeke, H.C. (1927). "Ein Walroß in der südlichen Nordsee". *Zoolog. Anzeiger*. LXXIV, (1). Leipzig: Akademische Verlagsgesellschaft, 89–90.

Ridley, G. (2005 [2004]). *Clara's Grand Tour. Travels with a Rhinoceros in Eighteenth-Century Europe*. London: Atlantic Books.

Ritvo, H. (1987). *The Animal Estate. The English and Other Creatures in the Victorian Age*. Cambridge, Massachusetts: Harvard University Press.

Ritvo, H. (2010). *Noble Cows & Hybrid Zebras. Essays on Animals & History*. Charlottesville and London: University of Virginia Press.

Robbins, L.E. (2002). *Elephant Slaves & Pampered Parrots. Exotic Animals in Eighteenth-Century Paris*. Baltimore and London: Johns Hopkins University Press.

Roberts, A. (1954 [1951]): *The Mammals of South Africa*. Johannesburg: "The Mammals of South Africa" Book Fund.

Rothschild, M. (1983). *Dear Lord Rothschild. Birds, Butterflies & History*. London: Hutchinson & Co.

Rothfels, N. (2002). *Savages and Beasts. The Birth of the Modern Zoo*. Baltimore and London: Johns Hopkins University Press.

Rothfels, N. (2008). "The Eyes of Elephants: Changing Perceptions". *Tidsskrift for kulturforskning* 7 (3), 39–50.

Rothfels, N. (2013). "Preserving History: Collecting and Displaying in Carl Akeley's *In Brightest Africa*". In: Thorsen, L.E, Rader, K., and Dodd, A. (eds.). *Animals on Display: The Creaturely in Museums, Zoos and Natural History*. University Park, Pennsylvania: Pennsylvania State University Press, 58–77.

Russel, G. W. (1927). "Walrus-Watching in Shetland." *Country Life* 15 January, 97–99.

Ryan, J.R. (2000). "'Hunting with the camera": photography, wildlife and colonialism in Africa." In: Philo C. and C. Wilbert (eds.). *Animals Spaces, Beastly Places*. London and New York: Routledge, 203–222.

Rydberg, G. (2008). "Banankungen Folke Anderson från Ödskölt – 40 år sen han mördades i Ecuador". *Hembygden*. Uddevalla: Dalslands Fornminnes- och Hembygdsförbund, 113–161.

Rydley, G. (2005). *Clara's Grand Tour. Travels with a Rhinoceros in Eighteenth- Century Europe*. London: Atlantic Books.

Sanderson, I.T. (1957). *The Monkey Kingdom. An Introduction to the Primates*. New York: Hanover House.

Setterberg, J. (1989). "Konservatorer som har skrivit historia vid Naturhistoriska museet i Göteborg". *Göteborgs Naturhistoriska Museum. Årstryck 1989*, 21–28.

Skottsberg, C. (1943). "Walter Kaudern in memoriam". *Särtryck ur Göteborg Museum Årstryck 1943*.

Snæbjörnsdóttir, B. and Wilson, M. (2006). *Nanoq: flat out and bluesome. A Cultural Life of Polar Bears*. Bristol: Black Dog Publishing.

Sorensen, J. (2009). *Ape*. London: Reaktion Books.

Star, S.L. (1992). "Craft vs. Commodity, Mess vs. Transcendence: How the Right Tool Became the Wrong One in the Case of Taxidermy and Natural History". In: Clarke, A.E and Fujimura, J.H. (eds.). *The Right Tools for the Job. At Work in Twentieth-Century Life Sciences*. Princeton: Princeton University Press, 257–287.

Stearn, W.T. (2001 [1981]). *The Natural History Museum at South Kensington*. London: The Natural History Museum.

Strong, Roy (1996). *Country Life 1897–1997. The English Arcadia*. London: Boxtree.

Sutcliffe, R., Rutherford, M. and Robinson, J. (2011). "Sir Roger the Elephant". In: S.J.M.M. Alberti (ed.), *The Afterlives of Animals*. Charlottesville and London: University of Virginia Press, 58–75.

Svanberg, I. (2001). *Siskeburar och guldfiskskålar. Ur sällskapsdjurens kulturhistoria*. Värnamo: Bokförlaget Arena.

Svanberg, I. (2007). "'Deras mistande rör mig så hierteligen". Linné och hans sällskapsdjur". *Särtryck ur Svenska Linnésällskapets Årsskrift*. Uppsala.

Svanberg, I. (2010). "'Nyss visades i Uppsala två små nyfödda lejonungar". Um svenska lejon". *Uppland*. Solna: Upplands Fornminnesförenings Förlag, 101–133.

Tagg, J. (1993 [1988]). *The Burden of Representation. Essays on Photographies and Histories*. Minneapolis: University of Minnesota Press.

Thackray, J. and Press, B. (2006 [2001]). *The Natural History Museum. Nature's Treasurehouse*. London: Natural History Museum.

Thamdrup, J. (1927a). "Hvalrossen paa Skagens Strand". *Dyrevennen* (1–2), 7.

Thamdrup, J. (1927b). "Hvalrossen paa Skagen". *Dyrevennen* (7–8), 42.

Thierry, B., Anderson J.R., Demaria, C., Desportes, C. and Petit, O. (1994). "Tonkean macaque behaviour from the perspective of the evolution of Sulawesi macaques". *Current Primatology*, 2: Social Development, Learning and Behaviour, 103–117.

Thorsen, L.E. (2009). "A Fatal Visit to Venice: The Transformation of an Indian Elephant". In: T. Holmberg (ed.), *Investigating Human/Animal Relations in Science, Culture and Work*. Skrifter från Centrum för genusvetenskap. Uppsala, 85–97.

Thorsen, L.E. (2009). "Speaking to the Eye. The Wild Boar from San Rossore." *Nordisk Museologi* (2), 55–88.

Thorsen, L.E. (2013). "Barry the Saint Bernard in Bern: A Dog of Myth and Matter". In: Thorsen, L.E, Rader, K. and Dodd, A. (eds.). *Animals on Display: The Creaturely in Museums, Zoos and Natural History*. University Park, Pennsylvania: Pennsylvania State University Press, 128–153.

Turchetto, M. (2004). *Morte di un elefante a Venezia: Dalla curiosità alla scienza*. Treviso: Canova.

Turner, A. (2013). *Taxidermy*. London: Thames & Hudson.

Wassén, H. (1942). "Walter Kaudern in Memoriam". *Etnologiska Studier 1941*, (12-13). Göteborg: Elanders boktryckeri, 305-330.

Wonders, K. (1993). "Habitat Dioramas. Illusions of Wilderness in Museums of Natural History". Acta Universitatis Upsaliensis. *Figura Nova Series 25*. Uppsala.

Yanni, C. (2005). *Nature's Museums. Victorian Science and the Architecture of Display*. New York: Princeton Architectural Press.

Østgaard Lund, H. and Berg, S.F. (eds.) (2011). *Norske polarheltbilder 1888–1928*. Oslo: Forlaget Press.

Index

INDEX